NOT SO FAST

NOT SO FAST

THINKING TWICE ABOUT TECHNOLOGY

DOUG HILL

THE UNIVERSITY OF
GEORGIA PRESS
ATHENS

Published by the University of Georgia Press
Athens, Georgia 30602
www.ugapress.org
© 2016 by Doug Hill
All rights reserved
Set in 10/13 Kepler Std Regular by Kaelin Chappell Broaddus
Printed and bound by Sheridan Books, Inc.
The paper in this book meets the guidelines for
permanence and durability of the Committee on
Production Guidelines for Book Longevity of the
Council on Library Resources.

Printed in the United States of America
20 19 18 17 16 C 5 4 3 2 1

Library of Congress Cataloging-in-Publication Data
Names: Hill, Doug, 1950–
Title: Not so fast : thinking twice about technology / Doug Hill.
Description: Athens : The University of Georgia Press, [2016] |
Includes bibliographical references and index.
Identifiers: LCCN 2016020562 | ISBN 9780820350295 (hard bound : alk. paper)
Subjects: LCSH: Technology—Social aspects.
Classification: LCC T14.5 .H55 2016 | DDC 303.48/3—dc23
LC record available at https://lccn.loc.gov/2016020562

CONTENTS

ACKNOWLEDGMENTS

Every page of this book demonstrates the debt I owe to the scholars whose work inspired it. They have been my guides for the intellectual adventure of a lifetime.

I'm grateful to Mick Gusinde-Duffy at the University of Georgia Press for believing in the book and for his unfailing enthusiasm and generosity of spirit throughout its gestation. Thank you, Mick. Thanks, too, to Jon Davies, Barbara Wojhoski, and everyone else at UGP who contributed to the process; they have been a pleasure to work with throughout. I'm also grateful to the scholars and writers who gave *Not So Fast* enthusiastic endorsements early on, before I found a publisher, among them Carl Mitcham, Jerry Mander, Albert Borgmann, Howard Rheingold, James Howard Kunstler, David Gill, Allen Noren, and especially Langdon Winner. Their validation was incredibly important to me personally and surely helped open the door to publication.

At an especially difficult time in my life, two friends recommended that the way out of the hole I was in was to take some time off to write the book on technology I'd been writing in my head for years. Thanks to Paul Geiger and Vincent Viglione for encouraging me to make the leap. Others who have been tremendously supportive throughout the long process of getting these thoughts into print include Peter Prichard, Paul Geiger (again), Roger Cubicciotti, Jim Landis, Dewar McCloud, Arielle Eckstut, Jerry Lazar, and Scott Hill. My wife, Judy Hill, had to share her home with a cloud of intrusive ideas for far too long. Thank you, dear.

The book is dedicated, with love, to Gram, Faith, and Jasper, who for better and for worse will live with the consequences of the technological world we've given them.

NOT SO FAST

I was introduced to what Martin Heidegger called "the question concerning technology" in 1992, when I moved from Manhattan to the leafy suburb of Montclair, New Jersey. The day we moved in I discovered that the house my then-wife and I had purchased was directly in the takeoff paths of two major airports, Newark and Teterboro. Wind patterns and timing had obscured that fact during our house-hunting expeditions. As a writer who works at home and who is especially sensitive to noise, I was devastated.

For a year or so I worked with the New Jersey Coalition against Aircraft Noise, a small but determined group of activists seeking to redirect planes out over the ocean instead of over densely populated communities. This was an option that would cost the airlines time and money, and they fought it. As the battle dragged on I began to understand that once technological systems are in place, they're very difficult to budge, and not only for political reasons.

Being far more inclined to reflection than activism, I soon found myself attending fewer hearings and buying more books. I discovered that scholars in two overlapping fields—the history of technology and the philosophy of technology—have spent their careers studying the influence of machines in our lives. I've been a voracious and admiring consumer of their erudition ever since.

This book is a combination of my absorption in the work of those scholars and of my own thinking about the issues they discuss. As my title suggests, I take a self-consciously skeptical view of mechanical miracle. I think it's too easy to take the promises of technology at face value and to ignore its hidden and not-so-hidden costs. Those assumptions need to be challenged.

This doesn't mean I don't appreciate the comforts technology provides. Like almost every writer these days, I spend much of my time sitting in front of a computer. I'm committed to air-conditioning and indoor plumbing. I spend a lot of time listening to my iPod, and I'm immensely grateful for the gift of Novocain.

I don't want to pass over this point facetiously. I recognize and acknowledge that technology's gifts go far beyond simple comforts, especially when linked with science. In my work as a health writer I recently interviewed an epidemiologist who is developing more-effective treatments for diseases that each year kill thousands of children who live in slums around the world. It's possible to be worried about overpopulation and still mourn the death of children. The concerns I address in this book are about the misuses of technology and about the temptations for misuse that accrue from the powers technology puts at our disposal.

When I first began studying the question concerning technology, there was almost no popular discussion on such topics as technological distraction, unexpected consequences, and the psychological effects of the Internet. That has changed, and I go out of my way here not to belabor arguments that have been made, often superbly, by others.

I try to follow the lead of one of my heroes, Lewis Mumford, who took what he called a "bird's-eye view" of our relationships to technology. Such a view can afford some crucial perspective that our daily immersion in a world of machines tends to obscure. The philosopher Albert Borgmann put it well: the true nature of technology can be understood only by looking at *the pervasiveness and consistency of its pattern.*" For that reason, case-by-case appraisals of individual devices or activities are necessarily incomplete. It is when we stand back to observe technological life in its "normal totality" (Borgmann's words) that we begin to see it for what it is.[1]

PART I

THE CLASSIC/ROMANTIC SPLIT

*Woe to that revolution which is
not guided by the historic sense.*

SAMUEL TAYLOR COLERIDGE

CHAPTER 1

THE PARADISE WITHIN THE REACH OF ALL MEN

These days you are to witness
Examples of my pleasing arts galore.
I'll give you what no man has seen before.

MEPHISTOPHELES TO FAUST

Let me begin by stating the obvious: We live in an era of technological enthusiasm.

It's not too vast a generalization to say that Americans, along with much of the world, are deeply, passionately in love with the technologies they use in their personal lives. We're also beguiled by the promises of scientists and engineers who say that, thanks to them, we'll soon be able to do just about anything we want to do. "At our current rate of technological growth," said Elon Musk, CEO of Tesla Motors and SpaceX, "humanity is on a path to be godlike in its capabilities."[1]

A similar trajectory has been offered by fellow enthusiast Marc Andreessen, cofounder of the breakthrough web browser Netscape, now one of Silicon Valley's leading venture capitalists. In a Twitter blitz in June 2014, Andreessen shared his vision of a future that is ours for the taking, if only we let the technologists work their magic without the constraints of regulation. When that happens, he wrote, we will enter "a consumer utopia" in which "everyone enjoys a standard of living that Kings and Popes could have only dreamed.... Without physical need constraints, we will be whoever we want to be."[2]

Often we're told it will take only a single technology to transform the human condition. "We're entering an age where the limits to our capabilities to re-make the world around us are limited only by our imaginations and our good judgment," proclaimed the reviewer of a book on synthetic biology. For Google chairman Eric Schmidt, the engine of our deliverance will be the Internet and the connectivity it provides. "If we get this right," he said during a conference presentation in 2012, "I believe that we can fix all the world's problems."[3]

Another remedy for all the world's problems had been predicted a few months earlier by Eric Anderson, copresident and cofounder of a company called Planetary Resources, which plans to mine precious metals from asteroids in outer space. Anderson is certain his project will produce unimaginable wealth, but that's just the beginning; it will also be the first step toward moving *all* industry into space, leaving nothing behind but verdant and peaceful landscapes. "We see the future of Earth as a Garden of Eden," he said.[4]

Such comments evoke a recurrent theme in the American experience: that we can cleanse all our past mistakes by opening a new frontier. Henry Ford had the same expectations for the slew of new technologies coming on the scene during his lifetime, which he said would deliver "a new world, a new heaven, and a new earth."[5]

Such comments also testify to a more recent wrinkle in utopian visions: that new technologies will be able to remedy the problems created by previous technologies. We see the same faith at work in the conviction of those who believe we'll come up with some way of reversing the catastrophe of global warming by "geoengineering" the climate of the entire planet. This is a sign that the technology enthusiasts of today are more aware than their predecessors that technology carries risks as well as promise.

For that reason their pronouncements, while still intoxicated and intoxicating, also tend to have disclaimers attached. A leading proselytizer of nanotechnology, Eric Drexler, for example, expects that within our lifetimes or those of our children, nano will place at our disposal a "genie machine" that will be able to assemble, molecule by molecule, pretty much any object we can imagine. "What you ask for, it will produce," Drexler has written, adding, however, that "Arabian legend and universal common sense suggest that we take the dangers of such engines of creation very seriously indeed."[6]

One of our more prominent and less restrained technology enthusiasts today is Ray Kurzweil, the inventor-turned-prophet who has captured a seemingly endless amount of media attention in recent years with his predictions of the imminent arrival—in 2045, to be exact—of "the Singularity." That's the historical turning point when humans will complete their ongoing merger with machines, creating a race of cyborgs with superpowers and without the annoying limitations of physical corporality.

"The Singularity will allow us to transcend the limitations of our biological bodies and brains," Kurzweil says. "We will gain power over our fates. Our mortality will be in our own hands. We will be able to live as long as we want (not necessarily forever). We will fully understand human thinking and will vastly extend and expand its reach."[7]

Among the gifts Kurzweil believes the Singularity will bestow:

- By the early 2030s, we'll be able to live happily and healthily without a host of body parts once considered vital. Bits soon to be obsolete include, Kurzweil says, the heart, lungs, stomach and lower esophagus, large and small intestine, bowel, red and white blood cells, platelets, pancreas, thyroid, kidneys, bladder, and liver. The functions of all will be taken over by a variety of techniques and devices, including synthetic, programmable blood; nanobots; various drug and dietary supplements; microscopic fuel cells; artificial hormones and intelligent biofeedback systems. Kurzweil suggests we may want to keep the mouth and upper esophagus because of the role they play in the enjoyable, though unnecessary, act of eating. With suitable improvements, we may also want to hang on to the skin, he said, given its important contribution to sex.[8]
- We'll be able to immerse ourselves in virtual realities without the bother of connecting to any devices, thanks to nanobots injected into our bloodstreams. These nanobots will interact with biological neurons to create "virtual reality from *within* the nervous system." As a result, Kurzweil says, we'll be able to imagine ourselves as being whomever we want to be, wherever we want to be, with whomever we want to be with, at any time.[9]
- Advances in digital processing will give us laptop computers that possess the intellectual power of "five trillion trillion human civilizations." As a result Kurzweil believes we will realize the Singularity's ultimate destiny: the entire universe will become "saturated with our intelligence."[10]

Although I'm not sure what saturating the universe with our intelligence entails, I'm less eager than Kurzweil to find out. As I see it, we've already saturated our own planet with our intelligence, and the results can be charitably described as mixed.

While it's true that the scale and scope of technological expectations have increased as the power and reach of technology itself have increased, the fact remains that utopia is utopia, whenever it's predicted. In that sense Kurzweil's expectations are entirely consistent with promises of technological deliverance we've been hearing for at least a couple of hundred years, a chorus of joyful proclamations that together amount to a venerable American tradition.

In 1853, for example, an anonymous author in the *United States Review* proclaimed that, thanks to technology, humankind's troubles would be ended within fifty years. "Men and women will then have no harassing cares, or laborious duties to fulfill. Machines will perform all work—automata will direct them. The only task of the human race will be to make love, study and be happy." Another author from the same period concluded, "Vanquished Nature yields! Her secrets are extorted. Art prevails! What monuments of genius, spirit, power!" A third asked, "Are not our inventors absolutely ushering in the very dawn of the millennium?"[11]

In *Technological Utopianism in American Culture*, historian Howard Segal reviewed twenty-five works of fiction published between 1883 and 1933, all offering visions of the glorious future technology would surely bring. Their authors shared several basic assumptions. Technological utopia was seen as not only possible but inevitable. The time and place of its arrival—within the next hundred years, usually; in the United States, always—could be accurately predicted, as could its characteristics. "This is a Utopian book," one author stated in his preface, "but its Utopia is not, as Utopias generally are said to be, in the clouds; on the contrary, it is worked out with much detail in accordance with a natural order of sequence from existing conditions, with every point definite in time and place, true in all fundamental physical features to the best maps, true also to the law of cause and effect and duly regarding the limitations of nature."[12]

Another similarity the authors of these books shared was a belief that, although the advance of technology would bring challenges of its own, in the end all difficulties would be surmounted by the power of technology itself. "They simply were confident," Segal writes, "that those problems were temporary and that advancing technology would solve mankind's major chronic problems, which they took to be material—scarcity, hunger, disease, war, and so forth. They assumed that technology would solve other, more recent and more psychological problems as well: nervousness, rudeness, aggression, crowding and social disorder, in particular. The growth and expansion of technology would bring utopia; and utopia would be a completely technological society, one run by and, in a sense, for technology."[13]

The first work of extended technological utopian thought to appear in the United States was John Adolphus Etzler's *The Paradise within the Reach of All Men, without Labor, by Powers of Nature and Machinery. An Address to All Intelligent Men*, published in 1836. Etzler was a German immigrant and peripatetic reformer; over the course of his career he spent time in Pennsylvania, the West Indies, and England and founded utopian communities in Ohio and Venezuela. His magnum opus proposed that the powers of the sun, tides, waves, and wind be harnessed for the benefit of humankind, and he wasn't shy about predicting, from his opening paragraph, the wonders his plans would bestow.[14]

"FELLOW MEN!" the book begins.

I promise to show the means of creating a paradise within ten years, where everything desirable for human life may be had by every man in superabundance, without labor, and without pay; where the whole face of nature shall be changed into the most beautiful forms, and man may live in the most magnificent palaces, in all imaginable refinements of luxury, and in the most delightful gardens; where he may accomplish, without labor, in one year, more than hitherto could be done in thousands of years . . . ; he may lead a life of continual happiness, of enjoyments unknown yet; he may free himself from almost all the evils that afflict mankind, except death, and even put death far beyond the common period of human life, and, finally, render it less afflicting: mankind may thus live in, and enjoy a new world far superior to our present, and raise themselves to a far higher scale of beings.[15]

Like many utopians, Etzler discussed the rewards he predicted in considerable detail, but there were gaps in his explanations of how, exactly, he planned to get there. For example, he proposed that a series of mile-long rows of sails, two hundred feet high, be erected on land. If those sails were adjusted by "mechanical contrivance" to accommodate shifts in wind direction, in a single twenty-four-hour day they would be able to produce "80,000 times as much work as all the men on earth could effect with their nerves." Specifics on the construction and maintenance of the mechanical contrivance or of the sails themselves were not provided, although Etzler did say that the two-hundred-foot height of the sails could be raised, if desired, "to the height of the clouds, by means of kites."[16]

To be fair, Etzler did provide elaborate mathematical calculations to back up some of his proposals; it's just that from today's perspective they don't seem very convincing. Not that they needed to be exact. Etzler insisted that even if his calculations were off, the scale of the powers ready to be exploited in nature would make any errors insignificant. Even so, he was aware that many would fail to take him seriously. "Studious" and "reflecting" minds would readily appreciate his proposals, he maintained. "But there will be also men who are so ill favored by nature, that they slovenly adhere to their accustomed narrow notions, without inquiring into the truth of new ideas, and will rather, in apology for their mental sloth, pride themselves in despising, disputing, and ridiculing what appears novel to them."[17]

However questionable Etzler's technical predictions may have been, he was correct on that score. One of those who had a hard time taking him seriously was Henry David Thoreau, who wrote, anonymously, a review of Etzler's book for the *United States Magazine and Democratic Review*. Titled "Paradise (to Be) Regained," it was slyly humorous in parts, openly sarcastic in others.[18]

"We confess that we have risen from reading this book with enlarged ideas, and grander conceptions of our duties in this world," Thoreau stated in his opening paragraph. ". . . It is worth attending to, if only that it entertains large questions." Parodying Etzler's enthusiasm, he continued, "Let us not succumb to nature. We will marshal the clouds and restrain the tempests; we will bottle up pestilent exhalations, we will probe for earthquakes, grub them up; and give vent to the dangerous gases; we will disembowel the volcano, and extract its poison, take its seed out. We will wash water, and warm fire, and cool ice, and underprop the earth. We will teach birds to fly, and fishes to swim, and ruminants to chew the cud. It is time we had looked into these things."[19]

Thoreau's basic point was that Etzler's energies were misdirected. Where the proposals in *The Paradise within the Reach of All Men* were aimed at reforming the Earth, he said, the man with a transcendental perspective aimed at re-forming himself. Etzler's schemes were therefore as unnecessary as they were grandiose. Thoreau granted that it was possible to imagine a future in which technological advance would make possible any number of improvements in everyday life, but he confessed such dreams left him uninspired.

"It is with a certain coldness and languor that we loiter about the actual and so called practical," he wrote. "How little do the most wonderful inventions of modern times detain us. They insult nature. Every machine, or particular application, seems a slight outrage against universal laws. How many fine inventions are there which do not clutter the ground?"[20]

One thing that's striking about the technology enthusiasts is how successfully they've managed, for centuries now, to ignore the reservations of skeptics like Thoreau. Their optimism is irrepressible, even though their promised utopia never seems to actually arrive. Yes, their predictions now come with disclaimers attached, but in comparison to the glories glimpsed just over the horizon, they inevitably sound pro forma.

Nevertheless, a quieter counterpoint of misgivings has always been there, testifying to a second great American tradition, the tradition of technological ambivalence. The tension between the poles represented by Etzler's enthusiasm and Thoreau's skepticism is as acute today as it ever has been, not surprisingly, given the powers our technologies have achieved and the scale of the dangers associated with those powers.

In fact, Etzler's ideas for using the sun, wind, and waves for virtually unlimited sources of energy are now being seriously considered by scientists who believe they may finally be capable of executing them. With any luck they'll figure them out in time to save the planet.[21]

CHAPTER 2

ABSOLUTE CONFIDENCE, MORE OR LESS

> Whatever the question, technology has
> typically been the ever-ready American answer,
> identified at once as the cause of the nation's
> problems and the surest solution to them.
>
> DAVID F. NOBLE

It's a truism that applies to most love affairs: once the initial euphoria has passed, a measure of ambivalence sets in. Certainly that's been the case with America's love affair with technology. Testimony to that fact has been the consistent appearance of representative figures, like Etzler and Thoreau, who personify, in terms appropriate to their times, our conflicted attitudes toward machinery, sometimes in opposition to other representative figures, other times within themselves.

Two individuals who have recently personified those parallel but opposing traditions in especially dramatic fashion are Steve Jobs and Ted Kaczynski, aka the Unabomber.

Some may be offended that I mention Kaczynski in the same breath as Jobs. Let me explain that the connection I'm talking about here isn't between Steve Jobs and Ted Kaczynski personally. Obviously there are huge differences between the lives of the two men. Rather I'm talking about them as archetypes, mirrors that reflect back to us our own feelings about technology.

The emotional reactions to Jobs's passing made it abundantly clear that for many of us he'd come to symbolize the hopeful, life-affirming potential of the technical arts, in the process buttressing our faith in technology as a vehicle of human progress. Kaczynski, by contrast, seemed a creature who'd emerged from the depths of our subconscious, a malignant manifestation of our fears that technology is not our friend but our enemy, and that our enemy is gaining the upper hand.

I'm known among my friends as a Luddite—the guy who can be counted on to grumble about how out of hand our national infatuation with technology has become—and I'm used to being considered something of an eccentric because of those views. For that reason I was surprised at the time of Kaczynski's arrest by the number of respectable people who expressed the opinion that, murders notwithstanding, his feelings about technology weren't entirely misplaced.

The journalist Robert Wright said in an essay for *Time* magazine that there's "a little bit of the Unabomber in all of us." An essay by Daniel J. Kevles in the *New Yorker* said the same thing, in almost the same words, under a headline reading "E Pluribus Unabomber."[1]

In his book *Harvard and the Unabomber*, Alston Chase argued at length that Kaczynski's manifesto was ignored not so much because its ideas were so foreign but because they were so familiar. Except for its call to violence, Chase believes, Kaczynski's message "embodied the conventional wisdom of the entire country. . . . It was nothing less than the American creed."[2]

Chase's book is a fine work overall, but that comment seems a huge overstatement. He goes on to cite a long list of popular books that represent the antitechnological consensus he feels exists, ranging from Al Gore's *Earth in the Balance* and E. F. Schumacher's *Small Is Beautiful* to Bill McKibben's *The End of Nature*. The same message shows up repeatedly, he adds, in contemporary children's stories and textbooks, reflecting the fact that Americans have long been "gripped by fear of, or revulsion against, the very technology the Unabomber now warned about: genetic engineering, pollution, pesticides, and herbicides, brainwashing of children by educators and consumers by advertising; mind control, cars, SUVs, power plants and power lines, radioactive waste; big government, big business; computer threats to privacy; materialism, television, cities, suburbs, cell phones, ozone depletion, global warming; and many other aspects of modern life."[3]

That's actually a more comprehensive list of technological worries than Kaczynski provided in his manifesto, but I doubt he'd quarrel with it. The question remains, though: Is Chase correct that these concerns place the manifesto squarely in the mainstream of American thought?

Polls suggest a somewhat different picture: enthusiasm overall with a strong undercurrent of unease. For example, the National Science Board's biennial surveys of American attitudes toward science and technology consistently show that a large majority of Americans—between 68 percent and 80 percent—believe that the benefits of scientific research outweigh its harms. At the same time, significant minorities of respondents have consistently agreed that "science makes our way of life change too fast" and that scientific research exceeds the bounds of what is "morally acceptable" (42 percent on both questions in the 2012 survey). A 2014 poll by the Pew Research Center found that 59 percent of Americans are optimistic that technological and scientific changes will make life in the future better, while 30 percent think these changes will lead to a future in which people are worse off than they are today. Respondents were less optimistic when asked about the potential benefit or harm of a range of specific technologies. More than half—65 percent—said we will be worse rather than better off if parents have the ability to produce smarter, healthier children by manipulating their DNA. Nearly identical percentages thought we'll be worse off if robots become the primary caregivers for the elderly and if drones are given permission to fly through most U.S. airspace.[4]

Other polls register similar results. The release of a 2012 Pew survey, for example, prompted this headline: "Americans Love (and Hate) Their Cell Phones." A 2000 poll by National Public Radio, the Kaiser Family Foundation, and Harvard's Kennedy School of Government found that Americans feel overwhelmingly positive about the digital technologies in their lives but also registered substantial concerns about privacy, pornography, exploitation of children, equal access to computers for the poor, and the impact of computer use on family life. A 2013 poll on public views regarding nanotechnology and synthetic biology found that most of those surveyed said they know little or nothing about either, although the more they learned about them, the more they worried about their risks. Nonetheless, a large majority of those surveyed continued to express confidence in the scientists and engineers who are exploring those technologies.[5]

These polls capture pretty accurately how Americans feel about technology, I think: ambivalent. And that strikes me as a perfectly logical position. Ted Kaczynski argued in his manifesto that we kid ourselves if we think we can separate good technologies from bad. That statement isn't as absurd as it may sound (a subject I'll get into later), but it also makes sense that some things about technology would please us and other things wouldn't. Kaczynski wrote contemptuously of Americans' inconsistencies toward technology: they want, he said, to have their cake and eat it too. To which Americans could reasonably reply, Sure we do. Who wouldn't?[6]

As I say, ambivalence toward technology is a long-standing American tradition. Thomas Jefferson set the pattern. In the earliest days of the republic the first American citizens actively debated whether to follow Britain in headlong pursuit of industrialization or to preserve the nation's agrarian character by continuing to base its economy on the family farm. Jefferson, believing that the American spirit drew its sustenance from nature, saw corruption and degeneracy as the inevitable by-products of large-scale industry. Dependence on manufactures "begets subservience and venality, suffocates the germ of virtue, and prepares fit tools for the designs of ambition," he wrote in 1781. "... The mobs of great cities add just so much to the support of pure government, as sores do to the strength of the human body."[7]

Jefferson was far from a primitivist, though, or a technophobe. He believed strongly in the importance of scientific discovery as the means by which the new republic could prosper and took pride in the newly established Patent Office as a sign of American inventive genius. During his travels abroad he collected and forwarded home news of the latest advances and could scarcely contain his excitement over the industrial power he witnessed in England. "I could write you volumes on the improvements which I find made and making here in the arts," he wrote to an American friend after visiting a steam-powered mill on the Thames. It's well known that Jefferson delighted in inventing and employing all sorts of gadgetry at home in Monticello, from a rotating turnstile in his clothes closet to an automatic door opener and an elaborate system of dumbwaiters. The Monticello compound also included a nail factory, a textile manufactory, and a gristmill, all employing the latest technology, all run by slaves, all supervised by Jefferson.[8]

These interests weren't incompatible with Jefferson's commitment to the family farm. He favored inventions that enhanced the practical abilities of the common man, including an efficient plow of his own design. Still, Jefferson found it necessary to significantly moderate his position on manufactures, deciding in the wake of the War of 1812 that unless the United States developed its own factories, it would forever be in thrall to foreign powers. "Experience has taught me that manufactures are now as necessary to our independence as to our comfort," he wrote in a letter in 1816. "... We must now place the manufacturer by the side of the agriculturist."[9]

Jefferson's phrasing there is significant. He thought it possible that industry and agriculture could peacefully coexist, side by side, a vision that historian Leo Marx has characterized as "the machine in the garden."[10]

Marx points out that contemporary standards of consistency can't be applied to Jefferson, who for decades claimed his sole ambition was to retire to the life of a gentleman farmer, even as he was pursuing one of the most stellar and peripatetic political careers imaginable. Jefferson himself wrote, in the same letter of 1816 quoted above, "no one axiom can be laid down as wise and expedient for all times and circumstances."[11]

This is not to say his shift on the subject of manufactures was made glibly. "It is important to stress his reluctance," Marx says. "One cannot read the long sequence of Jefferson's letters on the subject without recognizing the painful anxiety that this accommodation arouses in him." Jefferson's love of nature and his ambition for the fruits of industrial development represented a genuine split in his personality, one that is echoed in the ambivalence Americans have held toward technology ever since. Jefferson embodied, Marx believes, "decisive contradictions in our culture and in ourselves."[12]

At the same time it's also true that most of Jefferson's contemporaries believed scarcity was a greater threat to their virtue than factories and thus saw no reason to resist the lure of technological bounty. It's a mistake, said the renowned Harvard historian Perry Miller, to believe that industrialism somehow forced itself on a nation of pious, ascetic farmers.

> The truth is, the national mentality was not caught unawares, not at all so rudely jolted as is generally supposed. There were of course, as there still are, rural backwaters, where the people clung to the simpler economy and there was a certain amount of folk resistance to the temptations of the machine. But on the whole, the story is that the mind of the nation flung itself into the mighty prospect, dreamed for decades of comforts that we now take for granted, and positively lusted for the chance to yield itself to the gratifications of technology. The machine has not conquered us in some imperial manner against our will. On the contrary, we have wantonly prostrated ourselves before the engine.[13]

There's a touch of New England chauvinism in the presumption that those who resisted technological innovation—or, to put it another way, those who were reluctant to surrender traditional ways of life and livelihood—were confined to "rural backwaters." The same prejudice has long accounted for dismissive characterizations of the Luddites. Miller's comment also overlooks the degree to which proponents of technological expansion have successfully silenced, throughout American history, voices of opposition. Still, evidence supports the enthusiasm Miller describes.

Evidence also suggests, however, that undertones of anxiety have always haunted those fever dreams. Any number of historians, Perry Miller among them, have documented the profound unease that the Industrial Revolution created in a population that watched, bewildered, as certainties that had prevailed for centuries disappeared. America, Miller said, was caught up in a raging cataract of technological enthusiasm, only to find itself gradually swept out to a sea of disillusionment and self-doubt. And again, how could there not be ambivalence, given the waves of disruption that accompany each stage of technological advance? Rapid, radical change brings excitement and confusion at the same time. It's a recipe for ambivalence.[14]

Among those who sounded early alarms were some of the great names in American literature: Emerson, Thoreau, Melville, Hawthorne, Whitman, Twain, Poe, and Henry James. Many of them had their moments of technological optimism, Emerson, Whitman, and Twain in particular, but each was aware, eventually if not immediately, that technology removed as well as imparted. When he returned to his native country in 1904 after more than twenty years abroad, James was shocked by how America had changed, how crudely commercial and ugly it had become, and how passively Americans had acquiesced to that change. "The bullying railway orders them off their own decent avenue without a fear that they will 'stand up' to it," he wrote. He added that the environment had become so alien that he felt "amputated of half my history," and that he strained to hear "the last faint echo of a felicity forever gone." Gertrude Stein had a similar reaction when she returned, during a lecture tour in 1935, to her childhood home of Oakland, California: "There is no there there."[15]

Perry Miller warned that we should not assume that doubts expressed by literary elites reflected the feelings of the man and woman on the street. Prophecies of spiritual impoverishment were drowned out, he said, by the "background of loud hosannas" against which the prophets wrote.[16] I'm sure that's true, but it's also true that artists sometimes become memorable, or at least popular, not because they see what others fail to see but because their work manages to capture the spirit of their times. (This point comes up in a famous literary dispute I'll describe in the next chapter.) It was in a lecture titled "The Spirit of the Times," in fact, that Emerson bemoaned the selfishness and materialism of his era, pointing to technology as the cause.

> The new tools which we use, the steamboat, locomotive, and telegraph, are effecting such revolutions, as to induce new measures for every value, and to suggest a regret, which is daily expressed, that we who use them were not born a little later, when these agents, whose first machines we see prepared, should be in full play. We have been educated in stagecoach, spinning-wheel, and old tinder-box times, and we find ourselves compelled to accept new arts, new highways, new markets, and mend our old country-trot to keep up with these swifter-footed days.[17]

Misgivings about the spiritual costs of technology persisted even when the mythology of progress was arguably at its height, when the nation embarked on fulfillment of its Manifest Destiny. While most saw the settlement of the frontier as the triumph of civilization and prosperity over savagery and want, others saw natural beauty despoiled, natural virtue corrupted, human freedom eroded.

Historian Henry Nash Smith has documented how both perspectives were projected onto the personality of Daniel Boone. On the one hand, biographers pictured Boone as "the angelic Spirit of Enterprise," destined to bring commerce, wealth, and refinement to "the heathen wilds"; other accounts portrayed him constantly striving to stay a step ahead of the settling hordes. One newspaper article tracked Boone to Missouri after he'd fled Kentucky and Tennessee. "I had not been two years at the licks [in Missouri]," Boone was said to have complained, "before a d——d Yankee came, and settled down within a hundred miles of me!!"[18]

Boone is thought to have been one of several real-life models for the rugged woodsman Leatherstocking, hero of James Fenimore Cooper's hugely popular series of frontier novels, published between 1823 and 1841. Throughout the series Cooper wrestled thematically with the tension between respect for civilized order and the sacred freedom of untamed nature. The Leatherstocking persona lived on through later depictions of Kit Carson and Buffalo Bill and still later in the classic Westerns of film director John Ford. In each of these incarnations the self-reliant frontiersman acted as a bridge between wild nature and community, often demonstrating that for all the gains civilization brought, something noble and pure was being lost.[19]

The paradoxical ideals that were projected onto Daniel Boone also characterized the popular images of three of the most emblematic figures of the machine age: Thomas Edison, Charles Lindbergh, and Henry Ford. All were revered because of their unprecedented successes in furthering the technological project and at the same time beloved because they were thought to embody the homespun virtues of the all-American small-town boy.

There was a significant disjunction between Edison's public image as the lone genius and his role in establishing the corporate laboratories that would make the lone genius obsolete. One of his most influential contributions was the invention of the invention factory. It's also true that Edison's skill at designing complex technical systems was equaled by his facility for calculating the costs and potential profits of those systems and at raising the massive amounts of investment capital necessary to bring them to fruition. His backers included the likes of J. P. Morgan and the Vanderbilts, not the sort of company small-town boys usually keep.

Edison's success at gathering the big-time support he needed was due in part to his skillful manipulation in the popular press of his image as the straight-shooting, self-educated inventor from the sticks. Privately, Edison saw himself playing a quite different role: that of a baron of international industry. Thomas Parke Hughes and other historians believe his pursuit of that goal detracted from his focus as an inventor, contributing to his creative decline in later years.

Nonetheless, to the public Edison had it both ways. He proved that in a world increasingly dominated by industrial giants, the common man could still make a difference, in the process becoming an industrial giant himself. By the time of his death, Hughes says, Edison had achieved in the eyes of the public the status of a "secular saint" who represented what was best in the American character.[20]

For sheer intensity, even Edison's celebrity was eclipsed by the adulation accorded Charles Lindbergh. The hysteria produced by Lindbergh's solo flight to Paris in 1927 far outpaced anything Americans had experienced before and probably since. Any number of reasons were cited for this outpouring of emotion—Lindbergh's small-town origins, his modesty, the hunger of the nation for a hero at a moment of national malaise—but at the top of every list was the fact that this sweet-faced boy had been brave enough to face death all alone. Everywhere he was hailed as "the Lone Eagle," the solitary adventurer who had single-handedly conquered the Atlantic, much as Daniel Boone had conquered the wilds of Kentucky. Indeed, Theodore Roosevelt said that Lindbergh was Boone's "lineal descendent."[21]

Again, though, the public image was at odds with how the man saw himself. As historian John William Ward has pointed out, at every opportunity Lindbergh spoke of his flight as a collective experience. He didn't fly alone: he and his plane made the journey together, and the two of them were borne aloft by the expertise of the hundreds of scientists, technicians, and mechanics whose labors had advanced the progress of aviation. "Well, we've done it" were Lindbergh's first words when he climbed down from the cockpit in Paris; *We* was the title of the book he wrote describing the flight. Before he left France for America he was already scolding reporters for not saying enough about his plane's "wonderful motor"; he would subsequently call the *Spirit of St. Louis* "a living creature," as capable of loyalty as any human.[22]

John William Ward stresses that the public's worship of Lindbergh derived from the fact that it recognized in his flight the triumph of science and technology as well as the heroic achievement of a farm boy from Minnesota. This double-edged meaning amplified Lindbergh's celebrity by touching paradoxical emotions that lay hidden, for the most part, within the nation's collective unconscious. "The response to Lindbergh reveals that the American people were deeply torn between conflicting interpretations of their own experience," Ward says. "By calling Lindbergh a pioneer, the people could read into American history the necessity of turning back to the frontier past. Yet the people could also read American history in terms of progress into the industrial future. . . . The two views were contradictory but both were possible and both were present in the reaction to Lindbergh's flight."[23]

Contradiction marked the career of Henry Ford as well. As in the case of Lindbergh, the public's admiration for Ford stemmed from his combination of old-fashioned rectitude with machine-age achievement. Unlike Lindbergh, Ford seemed to embrace that contradiction without ever actually acknowledging that it was one.

Ford was one of the most recognized and respected public figures of the 1920s; Americans sent him thousands of personal letters each day. A group of college students named him the third greatest individual of all time, behind Napoleon and Jesus Christ, but he was also seen as the salt of the earth, a farm boy who never forgot where he came from. As the *New Republic* put it in 1923, "The average citizen sees Ford as a sort of enlarged crayon portrait of himself; the man able to fulfill his own suppressed desires, who has achieved enormous riches, fame and power without departing from the pioneer-and-homespun tradition."[24]

Ford shared that view of himself and actively promoted it. He was a tireless advocate of conservative Christian values and a relentless critic of the indecency of modern morals. Echoing Jefferson, he called the city a "pestiferous growth" and frequently compared the "unnatural" and "twisted" conditions of urban life to the purity and independence of the family farm. "When we all stand up and sing, 'My Country 'Tis of Thee,'" Ford wrote in an editorial, "we seldom think of the cities. Indeed, in that old national hymn there are no references to the city at all. It sings of rocks and rivers and hills—the great American Out-of-Doors. And that is really The Country. That is, the country is THE Country. The real United States lies outside the cities."[25]

Also like Jefferson, Ford envisioned a future in which technology and nature could peacefully coexist. "The new era will see a great distribution of industry back to the country," he said. "This country has got to live in the country; industry must be taken back to the country. . . . The great modern city is an abnormal development. It tends to break down under its own weight."[26]

Ford couldn't reverse the technological expansion he'd done so much to unleash, but he did have the resources to build a facsimile of what that expansion was about to displace. Disturbed that increasing road traffic and commercial development were threatening landmarks both personal and communal, Ford began spending millions to preserve selected specimens of authentic Americana, among them the Michigan farm where he'd grown up and a tavern in Massachusetts where George Washington and Lafayette had stayed. Eventually he focused his efforts on creating an entire preindustrial town. William Randolph Hearst had San Simeon; Henry Ford had Greenfield Village.

Greenfield Village occupies an expanse of more than two hundred acres not far from the Ford Motor Company's headquarters in Dearborn, Michigan. In his book *Ford, the Men and the Machine*, Robert Lacey describes Greenfield Village as "a never-never land as Norman Rockwell might have imagined it." There's a gristmill, a cider mill, gas lamps, and a horse-drawn omnibus taking passengers up and down Main Street. Some of the buildings stake claims to genuine historical significance: the courthouse where Abraham Lincoln practiced law, the bicycle shop where the Wright brothers built their first airplane, Edison's Menlo Park laboratory, the house where Stephen Foster was said to have composed "Swanee River." The intention, according to Ford's personal public relations man, was to recall "the real world of folks . . . that honest time when America was in the making."[27]

As Ford entered his sixties, Greenfield Village became his consuming passion. Where Hearst scoured the cathedrals of Europe to assemble a majestic castle, Ford scoured the small towns of America to re-create the humble surroundings of a simpler, slower place and time. During his travels he would often stop in an antique shop, peruse the items on offer, and instruct an aide to buy everything in it. Purchases were shipped to Dearborn, Lacey says, "by the railcar," everything from chandeliers and dressers to milk bottles, scrubbing boards, and flatirons. Mornings would find Ford at his vast River Rouge auto plant, one of the most advanced industrial factories in the world; afternoons would find him immersed in the sights, sounds, and smells of Greenfield Village, where he enjoyed taking meals in the kitchen of the farmhouse he grew up in. Ford not only reconstructed the past, Lacey says; he "reoccupied" it.[28]

Greenfield Village became a popular tourist attraction, with tour guides wearing old-timey costumes and old-timey artisans making old-timey crafts by hand. Next door Ford opened the Henry Ford Museum, which paid tribute to history of a different kind: the ever-advancing sweep of technological progress, complete with steam engines, wooden biplanes, and, of course, the Model T. An odd juxtaposition, perhaps, but one that also made perfect sense. As Robert Lacey put it, "Here were the Janus faces of Henry Ford—and of America herself—set together side by side."[29]

CHAPTER 3

AT THE INTERSECTION OF TECHNOLOGY STREET AND LIBERAL ARTS AVENUE

> Two souls, alas, are dwelling in my breast,
> And either would be severed from its brother;
> The one holds fast with joyous earthly lust
> Onto the world of man with organs clinging;
> The other soars impassioned from the dust,
> To realms of lofty forebears winging.
>
> FAUST

Ambivalence toward technology isn't limited to the American experience but has appeared repeatedly across the landscape of modern history, at least in the West, for centuries. This ambivalence strikes at something so central to the relationship between humans and technology that it can be extended outward to cultures as a whole and inward to the congenital dispositions of individual hearts and minds.

Again, Steve Jobs exemplifies the issues at stake, this time in relation to his longtime rival in the computer wars, Bill Gates.

Jobs often said that what distinguished Apple's products from those of its competitors was their combination of technology with human values. "It's in Apple's DNA that technology alone is not enough," he said during one of his famous product introductions. "We believe that it's technology married with the humanities that yields us the result that makes our heart sing."[1]

As he said this, a graphic appeared onscreen behind him showing a street sign depicting the intersection of "Technology" and "Liberal Arts." "I like that intersection," Jobs told his biographer, Walter Isaacson. ". . . The reason Apple resonates with people is that there's a deep current of humanity in our innovation."[2]

Nowhere was Jobs's conviction on this score more evident than in his contempt for the products produced under the aegis of Gates. Jobs regularly dismissed Gates as a man who traveled unswervingly on Technology Street, whizzing past Liberal Arts Avenue without a glance. Isaacson's biography rightly makes much of the tension between the two. Gates is described as "pragmatic" and "methodical" while Jobs is "intuitive" and "romantic." Gates had "abundant analytic processing power," while Jobs managed with "scattershot intensity and charisma." During face-to-face confrontations Jobs sometimes worked himself into rages or broke into tears, while Gates grew progressively cooler and more detached.[3]

Gates ultimately conceded that he admired Jobs's instinct for designing products people wanted to buy, even though he said Jobs "never knew much about technology." That's considerably more credit than Jobs was willing to give Gates, whom he dismissed as a "basically unimaginative" technician who might have brought more creative vision to his company if he'd dropped acid at some point, or maybe visited an ashram. "The only problem with Microsoft is that they have no taste, absolutely no taste," Jobs told Isaacson. "I don't mean in a small way. I mean that in a big way, in the sense that they don't think of original ideas and they don't bring much culture into their product."[4]

Quotes like these make me think of how familiar Jobs and Gates would have seemed to the great Romantic poet Samuel Taylor Coleridge. Coleridge wrote in 1860 that every man is born either a Platonist or an Aristotelian. By an Aristotelian he meant a person resolutely anchored to terra firma: practical and pragmatic, a calculator and a categorizer. It's an archetype that fits Bill Gates to a T. A Platonist, by contrast, is an idealist who believes there's more to the picture than meets the eye, who's convinced we live in a world of shadows that conceal invisible essences and that we sell ourselves short if we don't aim for higher truths. It's a good bet that when you hear someone talk about infusing technology with human values, you're listening to a Platonist.[5]

The insults exchanged by Jobs and Gates would also have been recognized by the philosopher and psychologist William James, who in a series of lectures in 1906 proposed the pragmatist as a mediator between the two types of personalities he said had long argued with each other in any number of fields: the "tough-minded" and the "tender-minded." James presented a list of characteristics that representatives of each orientation shared in common. Tough-minded people, he said, tend to be more empirical, materialistic, and skeptical, whereas the tender-minded are more idealistic and "monistic," meaning, I think, that they consider the day-to-day multiplicity of things as expressions of a single, higher truth. James proposed the pragmatist as a mediator between the two; such a mediator, he wrote, would have "scientific loyalty to facts" but also "the old confidence in human values and the resultant spontaneity, whether of the religious or romantic type."[6]

A more contemporary version of these contrasts can be found in a book that was very popular in the counterculture circles in which Jobs traveled as a youth, *Zen and the Art of Motorcycle Maintenance*. In all the interviews with him I've seen, Jobs never mentions having read it. It would be a shame if he didn't, because the intersection of Technology Street and Liberal Arts Avenue is its central theme.

Zen and the Art of Motorcycle Maintenance tells the story of a cross-country motorcycle trip that the author, Robert Pirsig, takes with his son, Chris, and two friends, John and Sylvia Sutherland. Along the way Pirsig engages in a series of philosophical reflections on the differences between what he calls the "romantic" and "classic" views of the world, views that correspond, more or less, to the Platonic and Aristotelian orientations defined by Coleridge.

A romantic understanding of the world, Pirsig said, "is primarily inspirational, imaginative, creative, intuitive. . . . It does not proceed by reason or by laws. It proceeds by feeling, intuition and esthetic conscience."[7]

The classic perspective is defined, by contrast, by rationality. Laws and reason are its guides. Its style is "straightforward, unadorned, unemotional, economical and carefully proportioned. Its purpose is not to inspire emotionally, but to bring order out of chaos and make the unknown known. It is not an esthetically free and natural style. It is esthetically restrained. Everything is under control. Its value is measured in terms of the skill with which this control is maintained."[8]

Pirsig's quest in *Zen and the Art of Motorcycle Maintenance* was to find a way of reconciling the tension between the classic and romantic points of view so that they might complement rather than oppose each other. It's regrettable, he believed, that people tend to see things in either classic or romantic terms and to view the other perspective with hostility and contempt.

Romantics, he said, tend to see the products of classic thinking as "awkward and ugly. . . . Everything is in terms of pieces and parts and components and relationships. Nothing is figured out until it's run through the computer a dozen times. Everything's got to be measured and proved." Classicists are no fonder of the romantics, whom they see as "frivolous, irrational, erratic, untrustworthy, interested primarily in pleasure-seeking. Shallow. Of no substance. Often a parasite who cannot or will not carry his own weight. A real drag on society."[9]

These descriptions are strikingly similar to the way Jobs once described his efforts to form an alliance between the technology industry and the "content providers" in the music and film industries. The people who run tech companies, he told Walter Isaacson, don't understand intuitive thinking. They think artists are lazy and undisciplined, while people in the music business are equally clueless about technology. "I'm one of the few people who understands how producing technology requires intuition and creativity," he said, "and how producing something artistic takes real discipline."[10]

Jobs was speaking there, I suspect, mainly of the executive types at those companies. Certainly he was aware that plenty of people besides himself have some affinity for both sides of the classic/romantic split. What made Jobs unusual was the scale of success he achieved by exercising those affinities. The stereotypes of technicians as human calculating machines and artists as drunken dreamers collapse pretty quickly under scrutiny. In practice, neither art nor technology are ever separate pursuits. Technicians can be as motivated as artists by the beauty of their work, just as artists are as interested as engineers in using tools and materials to achieve an effect.

Jobs said that many of the software and hardware engineers who worked with him on his breakthrough achievement, the Macintosh computer, were musicians and artists in their spare time. Most of us can probably identify elements of both tendencies in ourselves. Indeed, theories about the respective capabilities of the left and right hemispheres of the brain, or, more recently, Daniel Kahneman's slow brain and fast brain, suggest that the origin of these divisions may be found literally within each of us, in the firing of the neural networks that create consciousness. Still, coexistence doesn't necessarily mean peaceful coexistence, and the animosities between Jobs and Gates testify to an inherent oppositional tension between Platonic and Aristotelian personalities and pursuits.[11]

Jobs's contempt for the dreary lifelessness of Microsoft extended, to some degree, to his reservations about Western culture as a whole. This was a legacy of his sojourns in the counterculture and in India in particular. As he told Walter Isaacson, he found the Indians far more inclined toward "experiential wisdom" than intellectual rationalism. Jobs said that once he returned to the States he realized that the West's predilection for rational thought is a learned rather than an innate way of seeing the world, one that has produced great achievement but also substantial "craziness." Through his studies in Zen he worked at calming his mind in order to hear inner truths. "Intuition is a very powerful thing, more powerful than intellect in my opinion," he told Isaacson. "That's had a big impact on my work."[12]

With his focus on Buddhism and Zen, Jobs may not have been aware that the polarity between mysticism and rationalism has played a major role in the history of Christianity. The Eastern Orthodox tradition is more focused on contemplative spirituality than the Latin, Western tradition. It's partly for this reason that some cultural historians have argued that the more active spirituality of the Roman Catholic and Protestant faiths have paved the way for the nations of the West to emerge as the leading exploiters of technology. Given that the schism between the Greek and the Latin churches occurred in 1054, we can see that the tension between outer (rational/technological) and inner (intuitive/artistic) orientations to truth is a very old story.[13]

In fact, it's a story that precedes the Christian church by several centuries.

Coleridge's opposition of Plato and Aristotle would have us believe that the pattern was set in ancient Athens, and there's truth to that contention, though the differences between those pillars of inquiry can't be as neatly divided as Coleridge implied. The teachings of Plato and Aristotle overlapped to a considerable degree, not surprisingly, given that Aristotle was Plato's pupil. It's also true that neither philosopher articulated, based on the evidence that's survived, a consistent enough doctrine to say with assurance that Plato was a strict Platonist or that Aristotle was a strict Aristotelian.[14]

Overall, however, the Greek tradition was united in considering the pursuit of wisdom superior to engagement with the practical arts. This wasn't a strict rejection of everything we now consider technology or science. Plato, for example, considered mathematics, especially geometry, among the highest of wisdom pursuits. Always, though, universal understandings surpassed the production of artifacts. Meaning trumped utility.[15]

The Greeks distrusted technical skill because it threatened to bestow power on those least likely to use it responsibly. Classical Greek culture, writes the philosopher of technology Carl Mitcham, "was shot through with a distrust of the wealth and affluence that the *technai* or the arts could produce if not kept within strict limits." Allowed to flourish, the Greeks believed, technique led to excess and then to indolence, so that those who possessed it inevitably began to choose, as Mitcham put it, "the less over the more perfect, the lower over the higher, both for themselves and for others." This was a temptation that Steve Jobs, the perfectionist, famously resisted.[16]

Socrates considered farming the most virtuous form of manual labor, a line of thought that continued through Plato, Aristotle, Thomas Aquinas, all the way down to Thomas Jefferson and Wendell Berry. The ancients' mistrust of artifice also anticipated the conviction of Martin Heidegger, Jacques Ellul, and other modern philosophers that technique in its essence aims at transforming the natural world into means, rather than appreciating its intrinsic value as an end in and of itself. The Greek view that activity is inferior to contemplation, said Hannah Arendt, "rests on the conviction that no work of human hands can equal in beauty and truth the physical *kosmos*, which swings in itself in changeless eternity without any interference or assistance from outside, from man or god. This eternity discloses itself to mortal eyes only when all human movements and activities are at perfect rest."[17]

Jobs during his counterculture years would have appreciated this point of view; it's consistent with the perspective one acquires from the practice of Zen and also after the ingestion of LSD. In his business life, on the other hand, it's fair to say he departed from the Greek standard. Apple sells artfully designed objects of desire. Elegant they may be, but they're also extremely effective at distracting attention from the changeless eternity of the *kosmos*.

The Greeks' rejection of the practical arts was precisely the reason they

earned the enmity of Francis Bacon, who attacked them endlessly in his seminal works on the scientific method. Unlike the Greeks, the rationalism at the
heart of Bacon's program was from the outset directly linked to technology. The
goal was to produce *useful* results for the benefit of humankind. For that reason he despised the navel gazing of the ancient philosophers. Like children, he
said, "they are prone to talking, and incapable of generation, their wisdom being loquacious and unproductive of effects." The "real and legitimate goal of the
sciences," Bacon added, "is the endowment of human life with new inventions
and riches."[18]

That was a commission Steve Jobs certainly fulfilled.

CHAPTER 4

ANTIPATHIES OF THE MOST PUNGENT CHARACTER

> How insecure, how baseless in itself,
> Is the Philosophy whose sway depends
> On mere material instruments;—how weak
> Those arts, and high inventions, if unpropped
> By virtue.
>
> WILLIAM WORDSWORTH

William James noted, as Robert Pirsig did, that adherents on either side of the tough-minded/tender-minded divide tend to view each other with suspicion, if not outright hostility. In his series of lectures on pragmatism, James suggested that most members of his audience probably knew someone they'd identify as tough-minded and someone they'd consider tender-minded, "and you know what each example thinks of the example on the other side of the line.... The tough think of the tender as sentimentalists and soft-heads. The tender feel the tough to be unrefined, callous, or brutal.... Each type believes the other to be inferior to itself; but disdain in the one case is mingled with amusement, in the other it has a dash of fear." Throughout history, James added, these differences have produced "antipathies of the most pungent character."[1]

That would seem a fair description of the opinions Steve Jobs and Bill Gates had of each other, at least in the days when their rivalry was at its most heated. One of the more brilliant strategic moves in branding history had to be Apple's success at turning the "dash of fear" James described into a competitive advantage, portraying itself in TV ads as a bold or inventive underdog facing up to the clueless Microsoft and the totalitarian monolith IBM.[2]

Satire has long been employed as a weapon by those who feel themselves on the defensive, of course. Jonathan Swift used it in the seventeenth century to ridicule the legions of would-be scientists who set out to follow, worshipfully, Bacon's scientific method. Thoreau's tongue-in-cheek response to Etzler fits that pattern as well. By contrast, neither Etzler's *The Paradise within the Reach of All Men* nor Kaczynski's "Industrial Society and Its Future" (aka The Unabomber Manifesto) contain a shred of humor. Lewis Mumford has noted that a lack of levity is common among utopian writers, who take seriously the idea that perfection is attainable and therefore disregard the human folly that inspires satire.[3]

An irony of Kaczynski's terror campaign was the degree to which it was in sympathy with the values of the enemy he attacked. The evidence gathered at his wilderness cabin disclosed the obsessive planning of a dedicated technocrat. The measures he took to avoid discovery and capture, for example, included a four-grade system to classify the sensitivity of documents in his cabin and a ten-grade system to classify disposal methods for potentially incriminating waste materials. Detailed maps marked the locations of supply stashes hidden in the hills, should escape into the wilderness become necessary. Diagrams showed where in relation to the surrounding brush the stashes were buried and how deeply, measured to the inch; quantities and types of flour and other foodstuffs stored, calculated to the ounce; inventories of ammunition rounds, sorted by caliber, counted to the grain. His bombs were assembled, disassembled, and reassembled so obsessively that the FBI's profilers guessed the Unabomber might have gotten "some sort of bizarre sexual satisfaction" from the labor.[4]

Ted's brother David told the FBI that he and Ted had argued for years over the sufficiency of science versus the utility of art. David argued that there are "mystic unknowables" that can't be quantified or understood but that add immeasurably to the depth and meaning of existence. Ted—who had been, lest we forget, an assistant professor of mathematics at the University of California at Berkeley—would have none of it. He based his life, he said, on "the Verifiability Criterion." Facts are all that matter, and a fact is valid only insofar as it can be proved true or false.[5]

Kaczynski, as I've mentioned, was among a number of figures who have emerged in modern history to personify the tensions that result from our differing orientations to technology. I'll spend the remainder of this chapter describing two notable moments of conflict that similarly serve to illustrate our ongoing struggles with technological ambivalence. Both emerge, not by coincidence, at moments when the transformative powers of technology were especially ascendant.

The acceleration of the Industrial Revolution in the late nineteenth century was such a moment, setting the stage for an exchange between two of the more prominent figures of the Victorian era, the poet, educator, and social critic

Matthew Arnold and Thomas Henry Huxley, known as "Darwin's Bulldog" for his aggressive defense of the theory of evolution against its religious attackers.[6]

Huxley fired first, with an 1880 speech titled "Science and Culture." The occasion was the opening of Mason Science College in Birmingham, England, the epicenter of the Industrial Revolution. The school's founder, Josiah Mason, was a self-made, self-educated entrepreneur who'd made his fortune manufacturing jewelry, key rings, and pens—characteristic artifacts of that relatively recent arrival on the world stage, mass-market consumerism. Mason's bequest imposed on the college's administrators conditions that represented a substantial break from educational tradition, as well as a significant benchmark in the history of the classic/romantic split. Theology and "mere literary instruction" were to be excluded from the school's curriculum; students would focus on the natural sciences instead.[7]

Huxley heartily approved of these stipulations. The time had come, he believed, for a radical restructuring of higher education in England. Traditionally it was assumed that a proper English gentleman would immerse himself in the study of classic literature, specifically the philosophers of ancient Greece and Rome, read in the original languages, and the theology of the Christian church. Quoting Arnold's maxim that the meaning of culture was "to know the best that has been thought and said in the world," Huxley maintained that exposure to the ancients no longer satisfied that criterion.[8]

"For I hold very strongly by two convictions," he said. "The first is, that neither the discipline nor the subject-matter of classical education is of such direct value to the student of physical science as to justify the expenditure of valuable time upon either; and the second is, that for the purpose of attaining real culture, an exclusively scientific education is at least as effectual as an exclusively literary education." The ignorance of literary men regarding science and industry, Huxley added, was "almost comical."[9]

Arnold responded two years later with an address titled "Literature and Science," delivered as the prestigious Rede Lecture at Cambridge University. He acknowledged that the "crusade" to reverse the supremacy of literature over science in the traditional curriculum had gained "strong and increasing hold upon public favor." This did not mean, Arnold hastened to add, that he considered such a reversal desirable, or inevitable.[10]

Arnold insisted that his definition of what it means "to know the best that has been thought and said in the world" was far more capacious than Huxley assumed. To know the best that has been thought and said in the world referred not only to the giants of classic literature—*belles lettres*, as Arnold put it—but also to the works of *anyone*, ancient or modern, who had contributed significantly to our understanding of the world, including the works of the great natural scientists. "To know Italian *belles lettres* is not to know Italy," he said, "and to know English *belles lettres* is not to know England. Into knowing Italy and England there comes a great deal more, Galileo and Newton amongst it."

For Arnold the real question was one of emphasis. Some men, he said, would be satisfied to spend their lives collecting facts, whether those facts concerned the physiology of chickens' eggs or the grammatical structure of classical Greek. Darwin, who had died a few months earlier, might have been such a man, Arnold said, but his type was rare; they were "specialists." The interests of most men do not extend to the minutia of scientific processes, nor do they need to, Arnold said. For them, a general understanding of the discoveries of the natural sciences is enough.

What the vast majority of men *do* require, Arnold said—what the character of human nature requires—are four powers: "the power of conduct, the power of intellect and knowledge, the power of beauty, and the power of social life and manners." Moreover, human beings possess an innate hunger, an instinct, for relating or trying to relate these various powers to one another and for finding some way to link them into a coherent whole. That, Arnold insisted, was why focusing an educational curriculum on science at the expense of the humanities threatened a baleful narrowing of human understanding.

"Interesting, indeed, these results of science are, important they are, and we should all be acquainted with them," Arnold said.

> But what I now wish you to mark is, that we are still, when they are propounded to us and we receive them, we are still in the sphere of intellect and knowledge. And for the generality of men there will be found, I say, to arise, when they have duly taken in the proposition that their ancestor was "a hairy quadruped furnished with a tail and pointed ears, probably arboreal in his habits," there will be found to arise an invincible desire to relate this proposition to the sense within them for conduct and to the sense for beauty. But this the men of science will not do for us, and will hardly, even, profess to do. They will give us other pieces of knowledge, other facts, about other animals, and their ancestors, or about plants, or about stones, or about stars.... But still it will be knowledge only which they give us; knowledge not put for us into relation with our sense for conduct, [or our] sense for beauty, and touched with emotion by being so put; not thus put for us, and therefore, to the majority of mankind, after a certain while unsatisfying, wearying.

It wasn't an accident, Arnold added, that the ancient philosophers and theologians had been studied with such reverence for so long, for it was in their ability to make connections between our senses of knowledge, beauty, and conduct that they excelled. Their wisdom continued to carry profound meaning for us even after their knowledge of the physical universe had proved to be hopelessly ill-informed. Contrary to the assumption of materialists that the arts are "ornamental," elegant but superficial trifles to be pursued only as an afterthought, they are vital to our understanding of ourselves as human beings. The fact that the discoveries of the physical sciences were so effectively demolishing old assumptions about humankind's place in the universe made it more important, not less, that we remember the ancients and the continuities they revealed.

It's easy to recognize in Arnold's lecture the essence of the romantic response to rationalism, which is by definition a reactionary impulse, aimed at warding off a perceived assault. The gist of the problem is not rationalism per se but the tendency of rationalism to overwhelm everything else. As Ralph Waldo Emerson put it, "This invasion of Nature by Trade with its Money, its Credit, its Steam, its Railroad threatens to upset the balance of man & establish a new Universal Monarchy more tyrannical than Babylon or Rome. Very faint & few are the poets or men of God. Those who remain are so antagonistic to this tyranny that they appear mad or morbid & are treated as such."[11]

This same defensive impulse caused the mathematician and philosopher Alfred North Whitehead to declare his empathy for the Romantic poet William Wordsworth. "He felt something had been left out, and what had been left out comprised everything that was most important." Whitehead was one of those rare individuals who, like Steve Jobs, was able to achieve notable success on both sides of the classic/romantic divide. He described the Romantic movement as "a protest on behalf of value," a protest he explicitly joined. "We forget," he said, "how strained and paradoxical is the view of nature which modern science imposes on our thoughts."[12]

Huxley's attack on traditional education had been mostly unrelenting, but he did offer a concession on this score, insisting that the complete elimination of the classics was not his intention. "An exclusively scientific training will bring about a mental twist as surely as an exclusively literary training," he said. "The value of the cargo does not compensate for a ship's being out of trim; and I should be very sorry to think that the Scientific College would turn out none but lopsided men."

However strong the philosophic disagreements between Huxley and Arnold on the subject of education, their "antipathies of the most pungent character" were confined to their respective positions; they didn't extend to the personal. The two men were actually friends, which may have accounted for the relatively civil tone of their debate. Such was not the case with a more famous exchange on the same theme that unfolded three-quarters of a century later.

This time the fuse was lit by a lecture titled "The Two Cultures and the Scientific Revolution," delivered in 1959 by the British novelist C. P. Snow. Snow didn't mention in his lecture the Huxley-Arnold debate, an odd omission considering the occasion of its delivery—Snow spoke, as Arnold had, at Cambridge University's Rede Lecture—and the similarities in their subject matter.[13]

The two cultures of Snow's title were those of literary intellectuals, on the one hand, and of scientists, on the other. Snow, who had studied physics before turning to fiction, had friends in both camps, and he mourned the fact that "a gulf of mutual incomprehension" had developed between them. The literary experience of his scientist friends, he said, extended no further than Dickens, whom they considered esoteric, while his literary friends couldn't identify a scientific principle as basic as the Second Law of Thermodynamics. "Thirty years ago," Snow said, "the cultures had long ceased to speak to each other: but at least they managed a kind of frozen smile across the gulf. Now the politeness has gone and they just make faces." The result was a practical, intellectual, and creative loss to society as a whole.[14]

As he proceeded it became clear that the lion's share of Snow's disappointment was reserved for his literary friends, whom he described as uninterested in and contemptuous of science and technology. "Intellectuals, particularly literary intellectuals, are natural Luddites," he said. Snow argued that the intellectuals' insularity caused them to be insensitive to the suffering in the world, suffering that scientists and technologists were trying with their discoveries to relieve. Scientists have "the future in their bones," he said, while literary types act as if they "wished the future did not exist."[15]

The future, it turned out, was Snow's principal concern, specifically the future of what would come to be called the Third World. The poorer nations were well aware of the advantages the West had gained through industrialization, he said, and were determined to get those advantages for themselves. The only question was how peacefully or traumatically that goal would be achieved. Therefore the task of the West was to help developing nations get up to technological speed as quickly as possible, without paternalism.

Snow's shift of emphasis from the gulf between his literary and scientific friends to the necessity of spreading development worldwide was jarring, as Snow himself later acknowledged. He had originally thought of calling his lecture "The Rich and the Poor," he said, and in retrospect wished he had. That theme was overshadowed by the controversy over his comments on the two cultures, a result that caught Snow completely by surprise. He'd expected that after delivering the lecture he might receive a few letters in response and perhaps a comment or two in the press. Instead he watched as an avalanche of public discussion ensued, one that continued building momentum years after the event. Snow was certain that the scale of the response had far more to do with the timing of his remarks than with their originality; many before him, he said, had voiced similar opinions without attracting anywhere near the attention. It all served to convince Snow that the German philosophers of the nineteenth century had been onto something when they talked about the *Zeitgeist*.[16]

In his book *The Hedgehog, the Fox, and the Magister's Pox: Mending the Gap between Science and the Humanities*, Stephen Jay Gould suggests that Snow's lecture would have quickly faded from public view had it not been for a remarkably vehement counterattack that emerged from the literary quarter three years later. As the accounts of both Snow and his adversary suggest, this is incorrect. The original speech ignited its own controversy, which in turn occasioned the counterattack. No doubt the counterattack served to lift the controversy to new heights, as counterattacks often do, but it wasn't responsible for the uproar. Indeed, Snow's speech—the two cultures part of it—is widely remembered and referenced today, more than forty printings later. Few, however, remember the name of his opponent, F. R. Leavis, or what he said.[17]

Leavis was an author, critic, publisher, and longtime professor of literature at Downing College, Cambridge, where his lecture was delivered. (Marshall McLuhan was one of his students.) He was well known in English intellectual circles as a staunch defender of the unsurpassed sublimity of the great authors, whom Leavis saw as holding up an increasingly vital standard of excellence in the face of an onrushing tide of modern mediocrity. Snow represented to Leavis the perfect embodiment of that mediocrity and thus a clarion call to repel the barbarians at the gate.[18]

From his opening paragraph Leavis's attack was relentless. Snow's lecture displayed "an utter lack of intellectual distinction and an embarrassing vulgarity of style," its logic proceeding "with so extreme a naïveté of unconsciousness and irresponsibility that to call it a movement of thought is to flatter it." Leavis refused to call Snow a novelist—the quality of his literary output failed to justify the title, he said—and suspected his scientific credentials were equally thin.[19]

Leavis described taking a glance at a copy of Snow's remarks and deciding they weren't worth the three and sixpence it would have cost to purchase one. He then watched in horror as the lecture generated not only widespread comment but acclaim, then took on the aura of a modern classic. This, Leavis said, was why he felt compelled to react.

"Really distinguished minds are themselves, of course, *of* their age," Leavis said; "they are responsive at the deepest level to its peculiar strains and challenges: that is why they are able to be truly illuminating and prophetic and to influence the world positively and creatively. Snow's relation to his age is of a different kind; it is characterized not by insight and spiritual energy, but by blindness, unconsciousness and automatism. He doesn't know what he means, and doesn't know he doesn't know. . . . It's not the challenge he thinks of himself as uttering, but the challenge he *is*, that demands our attention."[20]

The acidity of Leavis's attack tended to overwhelm its substance. Beneath the invective was a passionate, if overtly elitist, romantic protest against the corrosive effects of science and technology.

He was especially incensed by Snow's comment that literary types tended

to be "natural Luddites" who had failed to grasp the significance of the Industrial Revolution. Such a statement ignored, he said, a century of distinguished
literary reaction to the moral and spiritual implications of the machine. Leavis
denied that he was a Luddite and insisted he had no desire to reverse scientific
progress or to deny the benefits it might bring to the developing world. The
point was that progress and development were not enough—"disastrously not
enough"—to achieve the felicity Snow envisioned.[21]

Leavis believed it more likely that such development would fatally under-
mine felicity in its truest, deepest form. Snow made the classic mistake of those
who saw salvation in industrial progress, Leavis said: he equated wealth with
well-being. The results of such a belief were on display for all to see in modern America: "the energy, the triumphant technology, the productivity, the high
standard of living and the life impoverishment—the human emptiness; emptiness and boredom craving alcohol—of one kind or another."[22]

If the rest of the civilized world was to avoid that fate, Leavis said, every re-
source of its "full humanity" would have to be mustered, together with "a basic
living deference" to a shared heritage that opens into an "unmeasurable" unknown to which each of us belongs.[23]

These comments, which appear near the end of Leavis's speech, come across
as strikingly poetic in comparison to the scorn that preceded them. They also
reveal the desperation that lurked beneath Leavis's anger. Clearly his bitterness
had been stoked by the vulnerability—the dash of fear—he felt as he faced an
enemy he recognized as not only unworthy but also overwhelming.

CHAPTER 5

A MOMENTARY INTERRUPTION

On what do you base your prediction that
the United States will disintegrate?

QUESTION ASKED OF MARSHALL McLUHAN
BY *PLAYBOY* MAGAZINE INTERVIEWER, 1969

It's often said that America is an optimistic country, and for most of our history, despite nagging misgivings, our attitudes toward technology have generally fit that description. Perry Miller was right: We've believed in the power of the machine to deliver us from darkness. The heyday of the sixties counterculture was an exception to that rule.

That Henry David Thoreau would become a hero of the hippies isn't surprising, for the idea that true reform comes from within was a key counterculture conviction and a key reason for its disenchantment with technology. The scale and depth of that disenchantment constituted one of the loudest and broadest reactions against technology in American history. It's no small irony that it was also the period that produced, thanks to Steve Jobs and many others, some of the most powerful technologies ever seen.

Several books have been written about the contributions of the counterculture to the computer revolution, and in general I admire them. Nonetheless, they tend to underplay, I think, the technophobic tenor of the time. It's true that the counterculture was a diverse, often scattered collection of subgroups (or "tribes," as the vernacular of the time put it) with widely differing visions and agendas. Among those subgroups were the technophiles who believed that computers, once placed in the hands of The People, could be potent forces for human liberation. The technophiles, however, were a counterculture minority—not a beleaguered, despised minority, but one that never occupied center stage. Their influence later was massively disproportionate to their influence at the time.[1]

This is not to say that the counterculture's attitudes regarding technology were entirely consistent. LSD was a product of a pharmaceutical laboratory, after all, and a high-quality sound system was an essential tool for unlocking the redemptive powers of rock and roll. Carl Mitcham has noted that, historically, romantic movements have evinced an "uneasiness" with technology that is "fundamentally ambivalent," meaning that romantics often share the ancient Greeks' skepticism toward technology but can also be seduced into the enthusiasms of the modern. And the counterculture was decidedly romantic.[2]

The fact remains, though, that the vast majority of the counterculture saw technology—advanced technologies such as computers and guided missiles in particular but also the mechanisms of mass consumption in general—as the enemy. The SDS's Port Huron Statement, for example, described automation as "the dark ascension of machine over man" and called technology as wielded by the existing power structure "a sinister threat to humanistic and rational enterprise."[3]

These attitudes go a long way toward explaining Steve Jobs's determination to bridge the arts-technology gap, and his understanding of why accomplishing such a feat was important. They also help explain where the seeds of Ted Kaczynski's revolution against technology were planted.[4]

One reason the counterculture's technophilic tendencies tend to be overplayed is the focus historians place on the influence of Stewart Brand's *Whole Earth Catalog*. Invariably the *Catalog* is cited as the embodiment of the counterculture's readiness to turn to tools of varying degrees of technological sophistication—from apple presses to lasers—in its quest for revolutionary change. The mistake, I think, is to cite the *Catalog*'s inclusion of high-tech products and ideas as representative of a receptiveness to those sorts of products and ideas on the part of the counterculture as a whole. It's more accurate to say that the presence of those products and ideas reflected the interests of Stewart Brand and other like-minded contributors. Steve Jobs and those like him responded to those interests because they shared them, but all sorts of people found the *Catalog* appealing for different reasons. It offered something for everyone. It was interesting to leaf through.

One author who got this right was Theodore Roszak, who wrote one of the touchstone books of the sixties, *The Making of a Counterculture*. Much later he published a witty essay on the counterculture's influence on the computer revolution, *From Satori to Silicon Valley*. In it he recalls, from the firsthand perspective of one who lived through it, the "voluntary primitivism" of a generation that strove, above all, to free itself from "the urban-industrial culture" that ruled the world.

Roszak tells a story about the first time he saw the *Whole Earth Catalog*. He was at a meeting somewhere on the San Francisco peninsula and somebody handed around the "rather ratty-looking" premier issue. Here's how he describes the reaction:

> It was closely scrutinized with a mixture of wide-eyed wonder and honest glee. For yes, here were the tools and skills of the alternative folk economy to come, the trial technology ready to be ordered and put to work. When the cities collapsed (as they were certain to do) and all the supply lines froze up (which might be any day now) these would be the means of cunning survival. Right there for all to see was a blueprint of the world's best teepee. There was even a book available for a modest price that showed how to deliver your own baby in a log cabin.[5]

Clearly, we're not talking here about a utopian vision of technological deliverance. We're talking about the means of survival in the wake of technological catastrophe.

Another author who got it right was Robert Pirsig. *Zen and the Art of Motorcycle Maintenance* was published in 1974, and its huge success was surprising in part because it challenged, or at least questioned, the technophobia of the generation responsible for that popularity. Yes, Pirsig seemed to be saying, I understand why you think technology is your enemy, but here's why that's not necessarily the case. At the same time he was fully aware of the ways in which the technological society had gone horribly wrong. His mission was to identify the errors of thinking and perception on both sides of the argument. He was searching for a way to heal the classic/romantic split.

Pirsig's traveling companions, John and Sylvia Sutherland, personified the romanticism of the counterculture. John, a professional drummer, was hip, which meant that he grooved rather than analyzed. Following instructions was for squares, and nobody in the sixties wanted to be a square. John and Sylvia rode from Minnesota to Wyoming without ever showing the slightest interest in learning how to maintain their bike. John reacted angrily, in fact, to any suggestion from Pirsig that maybe he should. (Sexual equality was not Pirsig's strong suit: the possibility that Sylvia might also have been the mechanic was never entertained.)[6]

As pathetic as John and Sylvia sometimes seemed, Pirsig recognized that their helplessness wasn't the only reason for their aversion to technology. Its ugliness and the damage it inflicts on the environment were part of it, but they were also responding to a deeper, more pervasive sense of alienation, a sort of lifeless mechanical essence exuded by the industrial landscapes around them. As they passed through those landscapes, they saw lots of machinery they didn't understand, lots of fences with "No Trespassing" signs on them, and lots of people moving around in circumstances that seemed overwhelmingly impersonal and depressing. It made them feel, Pirsig said, "like strangers in their own land."[7]

Steve Jobs would have recognized that allusion to another of the counterculture's touchstone texts, Robert A. Heinlein's *Stranger in a Strange Land.* He also would have understood the lifeless mechanical essence that so repulsed the Sutherlands. His sympathy with that view explained why a sense of friendliness and approachability would become key elements of his design philosophy at Apple.[8]

A little-known episode that occurred at the crest of the counterculture's influence speaks directly to both its predominate technophobia and its quieter dreams of technological salvation. It's worth recounting in some detail, I think, because the sixties were an era with grand ambitions and little appetite for compromise. Disagreements tended to be cast in the starkest of terms: you were either part of the problem or part of the solution. Beliefs for or against technology fit that pattern. It was a moment when the forces of enthusiasm encountered, for a change, an opposition with an equal or greater voice.

The episode in question concerns a flurry of excitement that surfaced in the mid-1970s over the possibility of establishing human settlements in outer space. The campaign's leading proponent was Gerard O'Neill, a physicist and professor at Princeton University. Two of his more fervent supporters were people whose influence would continue to be felt in technology circles for years to come: Eric Drexler and Stewart Brand.

Drexler was a graduate student at MIT who within the next decade would become one of the leading researchers in the burgeoning field of nanotechnology. Thanks to his 1986 book, *Engines of Creation: The Coming Age of Nanotechnology* (quoted in chapter 1), he would also become nanotech's leading proselytizer. Brand at this point was publishing a spinoff publication of the *Whole Earth Catalog* called the *CoEvolution Quarterly.* It might seem odd that a man so identified with the back-to-the-land movement was so eager to leave the land behind, but in fact Brand had long since decided that technology, not teepees, was where it was at. According to Andrew Kirk, author of *Counterculture Green: The Whole Earth Catalog and American Environmentalists*, Brand's early embrace of tools that would enable a retreat from industrialized society had given way to a growing affinity for and advocacy of more high-tech solutions. The problem was that many of his readers hadn't made that transition with him.[9]

O'Neill formulated his space-colonies idea at Princeton in the late sixties and gradually gained adherents over the next several years through a series of conferences, articles, interviews, lectures, and public forums. In July 1975 he testified before the House Subcommittee on Space Science and Applications and in January 1976 he appeared before the Senate Subcommittee on Aerospace Technology and National Needs. Brand heard him speak at a World Future Society conference in the spring of 1975 and was converted, he later wrote, from "mild interest in the Space Colonies to obsession." Drexler became a passionate supporter of the space-colonies concept after reading a 1974 article O'Neill wrote in *Physics Today*. He soon joined a group called the L5 Society, which dedicated itself to realizing O'Neill's dreams; the group's name was taken from the orbital address the first space colony would occupy. Drexler was a true believer; he told Brand, "I probably won't die on this planet."[10]

O'Neill's plan called for the construction of a series of permanently inhabited, self-supporting space colonies. Each colony would consist of gigantic rotating cylinders with attached appendages that would accommodate different areas for living quarters, light industry, heavy industry, and agriculture. The land area of one cylinder, O'Neill said, could be as large as one hundred square miles. Mirrors and shades could adjust ambient sunlight as needed to provide ideal conditions for each area. By varying the rotation of the cylinders, the level of gravity in different areas of the colony could also be adjusted "from zero to more than earth normal" and varied according to the needs in each area. Lower gravity in the area set aside for industrial operations, for example, would enable construction to be completed without the use of heavy cranes. An enclosed atmosphere would provide an oxygen level consistent with that at sea level on Earth. Because the colonies remained in permanent orbit, they would be able to take advantage of round-the-clock sunlight for the production of solar energy.[11]

O'Neill emphasized that the colonies' start-up costs would quickly be recouped through the sale of solar energy and of metals, mined first from the moon and then from asteroids, and that those operations would quickly make the colonies hugely profitable. The first colony could be established within fifteen years with a population of about 10,000, he said; from there inhabitants would increase steadily to about 250,000 by the year 2000.

O'Neill's presentations to Congress were dominated by charts, graphs, and diagrams, all clearly intended to make the project seem as practical and level-headed as possible. Nonetheless, plenty of specifics were left unexplained. He did promise that no breakthrough technologies were needed to make the space colonies a reality; essentially what we're talking about, he said, is "civil engineering on a large scale in a well-understood, highly predictable environment." Living conditions on the colonies would be comfortably familiar, with a few colorful exceptions. Abundant vegetation and animal life, lakes and streams suitable for swimming and boating, and hillside terraces would contribute to an environment similar to those of "some quite attractive modern communities in the U.S. and in southern France." Because levels of gravity could be varied, a short walk up a hillside could bring a resident to an area where "human-powered flight would be easy" and "sports and ballet could take on a new dimension."[12]

O'Neill kept glimpses such as these to a minimum, but Stewart Brand felt no need to be circumspect, at least at first. The fall 1975 issue of *CoEvolution Quarterly* devoted more than twenty pages to O'Neill's plan. Brand introduced the package with a glowing endorsement: "Space Colonies show promise of being able to solve, in order, the Energy Crisis, the Food Crisis, the Arms Race, and the Population Problem." Which was not to say they wouldn't also be fun.[13]

"Since the cylinders are big enough to have blue skies and weather," Brand wrote, "you might design a cylinder pair to have a Hawaiian climate in one and New England in the other, with the usual traffic of surf boards and skis between them (travel in Space is CHEAP—no gravity, no friction)." He also mentioned the human-powered flight feature, adding that the reduced gravity within some areas of the colony cylinders would allow you to dive into a swimming pool in slow motion. Space colonies, Brand concluded, were "readily possible— maybe inevitable—by 2000 AD."[14]

CoEvolution Quarterly's coverage included an article by Eric Drexler on the potential of mining asteroid belts. It contained the somewhat startling suggestion that the process might involve "sending out a work crew equipped with about one thousand 100 megaton hydrogen bombs." The bombs would be used to propel steel harvested from asteroids back to an orbit close to Earth, where it could be processed and then sold for billions of dollars profit. "If this proposal is to go it will need public and international acceptance of the detonation of hydrogen bombs in deep space," Drexler conceded. "This is, from physical grounds, an entirely safe thing to do because with the solar wind and the plasma environment of the solar system, one expects to receive, essentially, no materials of a radioactive nature or any other nature from the debris."[15]

A theme that surfaced repeatedly in Brand's thoughts about the colonies and often in comments from others was that they represented the opening of a new frontier. For Brand, outer space was "Free Space," an "outlaw area too big and dilute for national control." O'Neill, too, frequently employed the frontier motif. "The human race stands now on the threshold of a new frontier," he told the World Future Society, "whose richness surpasses a thousand fold that of the new western world of five hundred years ago." The book O'Neill subsequently published on his plan bore the title *The High Frontier*.[16]

The idea that space was a new frontier waiting to be exploited was one of the things that disturbed those who found O'Neill's ideas appalling. America had traveled that route before, they thought, with less than salutary results. Tens of thousands of native peoples had been murdered or exiled while a great wilderness was despoiled. Why should we think humankind would behave any more responsibly in outer space? The lines had been drawn for a war of words that would erupt in the next edition of *CoEvolution Quarterly*.

"Something about O'Neill's dream has cut deep," Brand wrote in his introduction to the debate. "Nothing we've run in *The CQ* has brought so much response or opinions so fierce and unpredictable and at times ambivalent."[17]

The reaction prompted Brand to solicit comments from his wide network of contacts and from his readership at large, and it was clear that he was surprised and to some extent chastened by the feedback he received. "It seems to be a paradigmatic question to ask if we should move massively into Space," his introduction to the published collection of letters continued. "In addressing that we're addressing our most fundamental conflicting perceptions of ourself [*sic*], of the planetary civilization we've got under way. . . . Is this the longed-for metamorphosis, our brilliant wings at last, or the most poisonous of panaceas?"[18]

Brand said in his introduction to the "Debate" issue that overall the reactions he'd received strongly favored the idea. That would change. In any event, it was hard not to be struck by the less favorable responses, both because they were so at odds with Brand's optimism and because of the names attached to them.

Lewis Mumford said that projects such as O'Neill's were typical of "technological disguises for infantile fantasies." Ken Kesey said he'd lost interest in "James Bond" ideas that abandoned the "juiciness" of life on Earth. E. F. Schumacher, author of *Small Is Beautiful*, offered to nominate five hundred people he'd be happy to send into space. Once they were gone, he said, the rest of us could get on with the work that truly needed to be done, which was developing technologies that would improve the lot of ordinary men and women.[19]

To be sure, there were positive responses. Buckminster Fuller, a particular hero of Brand's, said that for those attuned to "cosmic realism," the idea of living in space seemed the natural next step in the destiny of humankind. Astronomer Carl Sagan endorsed the frontier idea. Space colonies offered "a kind of America in the skies," he said, a place in which "the discontent cutting edge" could escape a world that had become "almost fully explored and culturally homogenized."[20]

The writer who emerged as the staunchest and angriest opponent of the space-colonies idea was a longtime friend of Stewart Brand and a longtime contributor to the *Whole Earth Catalog* as well as *CoEvolution Quarterly*: the novelist, poet, farmer, and activist Wendell Berry. His responses—and there would be several as the controversy unfolded—burned with a sense of righteous indignation, fueled not only by the implications of O'Neill's proposals but also by a sense of betrayal that Brand would so uncritically endorse them.

"Mr. Gerard O'Neill's space colony project is offered in the Fall 1975 *CoEvolution Quarterly* as the solution to virtually all the problems rising from the limitations of our earthly environment," Berry's first letter began.

> That it will solve all of these problems is a possibility that, even after reading the twenty-six pages devoted to it, one may legitimately doubt. What cannot be doubted is that the project is an ideal solution to the moral dilemma of all those in this society who cannot face the necessities of meaningful change. It is superbly attuned to the wishes of the corporation executives, bureaucrats, militarists, political operators, and scientific experts who are the chief beneficiaries of the forces that have produced our crisis.[21]

Although the space-colonies idea was perceived by its supporters as bold and innovative, Berry said, in truth it was an utterly conventional endorsement of the myth of progress, "with all its old lust for unrestrained expansion, its totalitarian concentrations of energy and wealth, its obliviousness to the concerns of character and community, its exclusive reliance on technical and economic criteria, its disinterest in consequence, its contempt for human value, its compulsive salesmanship."[22]

Berry was especially incensed by the idea that space was a new frontier. O'Neill and his supporters, he said, were only the latest in a long line of exploiters, from buffalo hunters to strip miners, who endorsed the myth that the ruin of one place can be corrected by hastening the ruin of another. Berry felt O'Neill wanted to become an inheritor of the frontier mentality without inheriting the tragedy of that mentality.

For Berry the question that needed to be addressed more than any other was that of restraint. He noted in particular Eric Drexler's proposal that a work crew be sent off to deep space bearing a thousand hydrogen bombs. The thought of it, Berry said, was nothing short of "monstrous."[23]

As I say, Brand was clearly stung by the negative responses the space-colonies issue received. He wrote a two-page editorial that tried to answer some of the major criticisms while maintaining his faith in the idea. Even if they fail, building the colonies will be important, he said, because then we will know, once and for all, that Earth is all we have. (He didn't say what would keep us from trying other space settlement options at some point in the future.) He insisted that space really *is* a new frontier because it really *is* unlimited, and as far as we know we won't be shoving any indigenous peoples aside when we go there.[24]

The defense that Brand seemed to find most appealing was that the space-colonies idea was *exciting*. It would stimulate ideas, discussion, and *movement*, especially among young people. He mentioned how many people attended *Star*

Trek conventions and how many read science fiction. Now was the time to get people who aren't engineers into the act, he said, including artists, novelists, poets, filmmakers, historians, and anthropologists—people "who can speak to the full vision of what's going on." In today's terminology, Brand was essentially calling for a crowd-sourced bridge of the classic/romantic split.[25]

Wendell Berry was having none of it. He wrote a second letter, longer and angrier than the first, more or less dismissing Brand's justifications as hopelessly naive. The only reason he was writing the second letter, he said, was that he intended to disassociate himself from *CoEvolution Quarterly*, and that the gentlemanly thing to do was to explain his reasons for doing so.

That prompted a gracious reply from Brand, again defending his support of the project (though with noticeably less conviction), declaring his affection and admiration for Berry, and urging him to reconsider. The mail he was receiving had turned dramatically against the space-colonies proposal, Brand said, so Berry shouldn't quit while he was ahead. "Besides," he added, "we've other fish to fry."[26]

Berry's response was also gracious but unbending. The stakes at issue were too important, he said, to shake hands and still be friends. He suggested he might reconsider if Brand would truly adopt the neutral editor's role he claimed he'd always intended to play in the debate, rather than serve as an advocate for O'Neill's "grandiose technological scheme."[27]

"I hope very much that you and I will have other fish to fry," Berry concluded. "But it's hard to have an appetite for fish when you've already got a bone stuck in your throat."[28]

The space-colonies project soon lost momentum and disappeared from public discussion. Stewart Brand moved on to other heroes, Nicholas Negroponte of MIT's Media Lab among them, and threw his remarkable energy and force of personality into the computer revolution. His list of credits there includes editor and publisher of the (short-lived) *Whole Earth Software Catalog* and cofounder of one of the first and most influential cyber communities, the WELL (an acronym for Whole Earth 'Lectronic Link). He also helped organize the first national conference for computer hackers and was a charter member of the board of directors of the Electronic Frontier Foundation, a nonprofit lobbying group dedicated to protecting freedom of speech, privacy, and other "digital rights" on the Internet.

The name of the Electronic Frontier Foundation demonstrates that the frontier mythology never dies; it just shifts to the latest technology. And indeed, the pull of the frontier remains a potent driver of dreams today, especially for tech tycoons with money to burn. Google's Larry Page and Eric Schmidt, for example, are investors in Planetary Resources, the asteroid-mining project mentioned in chapter 1. Elon Musk and Amazon.com's Jeff Bezos have founded rocket companies of their own. Bezos talks of establishing "an enduring human presence in space"; Musk plans to build a civilization on Mars.[29]

Another example is Peter Thiel, cofounder of PayPal and a key provider of early seed money for Facebook. Thiel told one reporter he spends what may be an unhealthy amount of time thinking about frontiers. He was the largest single funder of Ray Kurzweil's Singularity University and is a major supporter of the Seasteading Institute, which was founded to establish floating cities in international waters. These island communities were envisioned as libertarian laboratories. Residents would encounter no encumbering laws, no obstructive government. Just freedom. They promise to be, as Thiel himself put it, an opportunity to "go back to the beginning of things," a place where human beings can "start over."[30]

Back, in other words, to the Garden of Eden.

PART II

THE WATER WE SWIM IN

*Did you ever look at something
and it's crazy, but if you
look at it in another way
it's not crazy at all?*

ROY IN
CLOSE ENCOUNTERS
OF THE THIRD KIND

CHAPTER 6

WHAT IS TECHNOLOGY?

> All descriptions or definitions of technique which
> place it outside ourselves hide us from what it is.
>
> GEORGE GRANT

There's a story about an older fish passing two younger fish in the ocean. "How's the water?" the older fish asks.

The young fish look confused. "What's water?" one asks.

Our ability to answer the question, "What is technology?" would likely elicit a similarly clueless response, for the same reason. We're so immersed in it we take it for granted without ever pausing to think about what it is.[1]

Witness the almost automatic assumption today that "technology" refers to *digital* technologies, especially the Internet and smartphones, and maybe certain high-tech machines like satellites and particle accelerators. This is a definition that excludes a universe of everyday technological artifacts, everything from cars and can openers to toilets and tricycles—all things that most of us would quickly agree fit under the rubric "technology" if we thought about it, but we don't.

This is not to say that technology is easy to define, even when you do think about it. Indeed, even the historians and philosophers who have made the study of technology their life's work have failed to settle on a definition that satisfies everyone. "Technology can be defined no more easily than politics," writes Thomas Parke Hughes, one of the leading historians of technology. "Rarely do we ask for a definition of politics. To ask for *the* definition of technology is to be equally innocent of complex reality."[2]

Nonetheless, definitions abound. Hughes himself offers this: "The effort to organize the world for problem solving so that goods and services can be invented, developed, produced, and used." Carl Mitcham, one of the leading figures in the philosophy of technology, is briefer: technology, he says, is "the human making and using of material artifacts in all forms and aspects." Edward Tenner, the Princeton professor who helped popularize the phrase "unintended consequences," is briefer still: he calls technology "the human modification of the natural world."[3]

Definitions of technology sometimes carry implications hidden to those not attuned to an argument in progress. A case in point is that of Joseph Pitt, a professor in the philosophy and history of technology at Virginia Tech. Pitt, a leading figure in the field, is one of those who has repeatedly challenged what he sees as the antitechnology bias of many of his colleagues. Technology, Pitt says, is "humanity at work." This work, he adds, constitutes an ongoing endeavor that involves "the deliberate design and manufacture of the means to manipulate the environment to meet humanity's changing needs and goals."[4]

Pitt's use of the word "deliberate" is, I suspect, a deliberate way of making the point that it is humans who use tools, not the other way around. Pitt is especially annoyed by those who, in his opinion, reify technology, causing them to conclude that it somehow takes on a life and will of its own. No doubt Pitt's number one suspect on that score would be Jacques Ellul.

A French sociologist and theologian who died in 1994, Ellul has attracted a small but passionate body of admirers over the years. Among scholars, however, he remains something of a fringe figure, largely because his belief that technology plays a formative role in human culture is out of step with academic fashion. Although he usually kept his theological writings separate from his writings on technology, Ellul's religiosity was also out of step with academic fashion, as was his unrestrained style of argument—his writing is filled with colorful description, irony, and righteous anger. Nonetheless, his insights into the character of the technological phenomenon make him hard to ignore. There are those who consider Ellul a genius. I'm one of them.[5]

In *The Technological Society* and other books, Ellul argued that we're mistaken in thinking of technology as simply a bunch of different machines. In truth, he said, technology should be seen as a unified entity, an overwhelming force that has already escaped our control. Ellul used the word "technique" to underscore his conviction that technology is not only hardware, but also a state of being that incorporates everything it touches, including humans, into itself. Machines are "deeply symptomatic" of technique, he said. They are "the ideal toward which technique strives."[6]

Ellul is far from the only philosopher to see technology as more than tools or machines. In fact, he's on one side of a fundamental split in the field between those who define technology broadly and those who define it narrowly. Joseph Pitt uses the words "internalist" and "externalist" to define the same distinction. The narrow, internalist school holds that technology means tools, period. That's oversimplifying, but the point is to keep the question focused on objects, artifacts, hardware: *things*. People with engineering backgrounds, like Pitt, tend to favor that perspective.[7]

Those who favor the broad or externalist view tend to have more sociological or philosophical backgrounds and tend to see technology as a *system*, one that includes not only the machines but their users and the broader social and political contexts in which they're used. As Thomas Misa, a history of science and technology professor at the University of Minnesota, put it, "Technology is far more than a piece of hardware. Properly understood, 'technology' is a shorthand term for the elaborate sociotechnical networks that span society. To invoke 'technology,' on the macro level of analysis, is to compact into one tidy term a whole host of actors, machines, institutions, and social relations."[8]

By no means is the systems perspective shared only by academics, as evidenced by a comment from Simon Ramo, cofounder of TRW Corp., a veteran of Hughes Aircraft and General Electric, and a graduate of Caltech. "Systems engineering is inherently inter-disciplinary," he said, "because its function is to integrate the specialized separate pieces of a complex of apparatus and people— the system—into a harmonious ensemble that optimally achieves the desired end."[9]

Technological systems have a very long history. Lewis Mumford, one of the best-known philosophers and historians of technology, argued that the Egyptian pyramids were one of the earliest examples of what he called "the megamachine." The defining characteristics of the megamachine, he said, are "a multitude of uniform, specialized, interchangeable but functionally differentiated parts, rigorously marshaled together and coordinated in a process centrally organized and centrally directed: each part behaving as a mechanical component of the mechanized whole." The main difference between a megamachine that builds pyramids and one that, say, bottles soft drinks is that the pyramid-building machine used mostly human parts; it was, Mumford said, a "thousand-legged human machine." Mumford called the megamachine "the supreme feat of early civilization: a technological exploit which served as a model for all later forms of mechanical organization."[10]

Although the defining characteristics of the megamachine may have been present for centuries, it's also true that technology as a systemic phenomenon has certainly become steadily more prevalent, and more pervasive, since the Industrial Revolution. As Ellul often noted, the growth of technique is cumulative, a point with which many technology enthusiasts would agree.[11]

My own view of technology aligns with the likes of Mumford and Ellul, but it's only fair to acknowledge the objections of those who feel otherwise. Joseph Pitt fills the role of antagonist admirably, especially in his complaints that Ellul and others fall into the trap of reifying technology and then attributing causal powers to it. As Pitt sees it, this makes it all too easy to blame technology for problems associated with technology, when in truth it is always human beings who make decisions regarding how, where, and when technologies are employed. Our fear of reified technology is a diversion aimed at avoiding our own responsibility.

"It is not the machine that is frightening," Pitt says, "but what some men will do with the machine; or, given the machine, what we fail to do by way of assessment and planning." Pinning the blame on machines rather than humans, he adds, is a form of "intellectual hysteria" that "makes successful dealings with the real world impossible."[12]

There's certainly truth to Pitt's characterization, but it's crucial to acknowledge that technology, especially when we consider technology as a system, is not as readily controlled as he suggests. Technology is a form of embodiment that becomes a presence and a phenomenon with often-inflexible characteristics. While it's true that communities can alter those characteristics if they choose to, it's also true that it is often practically impossible to do so, simply because the associated costs, social as well as economic, are prohibitive. Contrary to Pitt's assertion, technology is not infinitely malleable. I will examine the depth and dynamics of that inflexibility in the next two chapters.

Pitt is also mistaken if he believes that taking a broad, systemic view of technology automatically implies a surrender of human responsibility. Jacques Ellul, for example, argued that the powers of technique work to neutralize the ability of human beings to put up an effective resistance, should they wish to, but he didn't believe this happens without a degree of human participation and consent. Note, for example, this passage from one of his most important theological works:

> Up to the eighteenth century man was obviously in control of his technical life. But once technology reached a certain perfection and dimension and the technical phenomenon arose, technology threw off completely the control of man and became in some sense the destiny or even the fate of man. It should be noted, however, that this could happen only because there was not merely an inner mechanism but because man consented to go along with it. Man is not stripped of his mastery apart from or contrary to himself. The movement of reversal always has two aspects, first, the expansion of the phenomenon by its own inner dynamic, and secondly the psycho-spiritual attitude of man, which can take such different forms as total unawareness, simple passivity, rational assent, mythical hope and delirious passion.[13]

Lewis Mumford was similarly careful to guard against the impression that the megamachine could operate without benefit of human agency. Technique, he said, "does not form an independent system, like the universe; it exists as an element in human culture and it promises well or ill as the social groups that exploit it promise well or ill. The machine itself makes no demands and holds out no promises; it is the human spirit that makes demands and keeps promises."[14]

That goes too far, I think. Machines *do* make demands and hold out promises, simply because demands and promises are built into machines by the human beings who design and implement them.[15]

Some observers have seen the evolution of technology as gradually leaving the machine itself behind. Thus, we're told, the "information society" depends on "soft" technologies that are increasingly malleable, eventually devolving into bits of digital data. The suggestion is that there's something unique about digital technologies, that they're taking us into an entirely different era.

My own view is that this contention is false from one perspective and true from another. On the one hand, it's important to recognize that there are underlying commonalities to all technologies. Thus, we're mistaken if we think of digital technologies as somehow representing a radical break from their industrial and preindustrial predecessors. It's often said, for example, that the Internet is a technology that connects people with one another, thereby enhancing human community. In the nineteenth century exactly the same thing was said about railroads.

On the other hand, it's also true that technological advance is cumulative. This evokes a principle articulated by Hegel and often repeated by Ellul: that at some point a change of quantity becomes a change of quality. This is especially true if you agree that "quantity" connotes potency as well as ubiquity. A fundamental quality of technological power is action at a distance. In that sense the technological world we live in today is qualitatively different from that of a century ago, or half a century ago.[16]

One of the trickier definitional issues to be clarified is the relationship between science and technology. Which is the parent and which is the child? This was an odd lacuna in Ted Kaczynski's thinking: he revered science but hated technology, never seeming to consider how closely the two are intertwined.[17]

For a long time the assumption was that science begets technology, a view reflected in the motto of the 1933 World's Fair: "Science Finds, Industry Applies, Man Conforms." By then scholars had already begun to argue that, if anything, science was applied technology. As the scientist/sociologist/historian L. J. Henderson put it, "Science owes more to the steam engine than the steam engine owes to science." In the mid-1960s Yale professor Derek J. de Solla Price published a series of papers supporting that view. Among the examples he cited was Galileo's use of the telescope to affirm the Copernican Revolution, a breakthrough that depended not only on the technology of the telescope itself but on the technology of the lens grinders who made the telescope possible. "Much more often than is commonly believed," Price wrote, "the experimenter's craft is the force that moves science forward."[18]

Another dimension was added to the discussion in 1970 by Cyril Stanley Smith, a professor emeritus at MIT who cited a host of examples documenting that the creation of many technologies in their earliest incarnations was driven as much by the artistic impulse as anything else. Ceramic figurines preceded ceramic pots, for example, while copper was used to make beads before it was used to make swords. Curiosity about the use of explosives for purposes of war was ignited by the use of fireworks for purposes of delight, while the block printing of fabrics helped establish techniques later used in movable type. "Over and over again," Smith concluded, "scientifically important properties of matter and technologically important ways of making and using them have been discovered or developed in an environment which suggests the dominance of aesthetic motivation."[19]

The title of Smith's article—"Art, Technology, and Science: Notes on Their Historical Interaction"—underscores the fundamental point that any attempt to draw firm boundaries between those respective fields is fruitless. The most widely accepted view of the relationship between technology and science today, in fact, is the "interactive" model, which holds that the two disciplines comprise independent cultures that sometimes advance on their own and at other times draw on each other. Edwin Layton Jr., a leading historian of technology, has written that science and technology "are best seen as the products of their mutual co-evolution, at least in the modern era. That is, the two really are symbiotic with one another. Neither is the exclusive property of a single community." Otto Mayr, another prominent historian of technology, added that confusion between the two realms is inevitable, given the broadness of their scope and the degree to which they naturally overlap. "The words 'science' and 'technology' are useful precisely because they serve as vague umbrella terms that roughly and impressionistically suggest general areas of meaning without precisely defining their limits," Mayr said. The motives driving both endeavors, he added, are similarly mixed and complex.[20]

Typically, Jacques Ellul's thoughts on the subject were less measured. Ellul called the idea that technology is applied science "radically false." Technique, he said, has been a fact of human existence since prehistory but didn't really take off until science caught up. At that point technology exploded, but science has always been dependent on technique, and has become more so over time. "Today, all scientific research presupposes enormous technical preparation," Ellul wrote. "When the technical means do not exist, science does not advance."[21]

An overstatement? Perhaps. Still, the predominance of evidence suggests Ellul got it mostly right. One thinks of the frustrations of the world's physicists as they waited years for the Large Hadron Collider to get up and running, or the excitement of NASA's chemists as they waited for their robot rover to send back data on the soil samples it was collecting on Mars, or the insights neurobiologists are gaining thanks to computed tomography scans of the human brain. Many of the hottest fields in science today—genetics, nanotechnology, and artificial intelligence among them—depend almost entirely on technology.

The job title of Stanford University's Drew Endy, a leader in the field of synthetic biology, tells the story: he's an assistant professor of "bioengineering." Endy says it's no mystery how we managed to go from a standing start to sequencing the human genome in less than a decade. "It's not because [prominent leaders in the field such as] George Church and Craig Venter and Eric Lander and Francis Collins got ten billion times smarter during the Clinton years," he said. "It's because the technology for sequencing DNA got automated and scaled up sufficiently to do it."[22]

Although it's often said that science aims at knowing and technology at making, more and more the two pursuits have merged, thanks to the growing pressure on scientists to justify the costs of research by producing a marketable product. This is something else Ellul predicted. Increasingly science has become, he said, "subordinate to the search for technical application."[23]

This raises an even trickier definitional question: how to decide which partner leads in the dance between technology and capitalism. Without question an intimate relationship has evolved between them, as Marx prophetically recognized. There's also no question that technology preceded capitalism, as Mumford's megamachine makes clear. Ellul's use of the word "technique" tends to blur the line between the two; if pressed I think he would have argued that capitalism *is* a technique.

Certainly capitalism has proved to be the economic system best suited to the unhindered development of technology in all its forms. Lenin and Stalin tried their best to exploit the rapidly growing technological powers of their era (like Hitler, they admired in particular the assembly lines of Henry Ford), but American-style consumerism eclipsed Soviet communism, just as it now aims to eclipse Chinese communism and Islamic theocracy. The fact that free-market capitalism has for all intents and purposes replaced democracy as the American form of government testifies to the transformative capacities of the technologies at its disposal.

The historian David F. Noble has argued that technology is "the racing heart of corporate capitalism," implying that capitalism directs the enterprise while technology supplies the motive force. I think you could just as successfully argue that technique directs the enterprise while capitalism supplies the motive force. As Alfred D. Chandler put it in the introduction to his seminal work of economic history, *The Visible Hand*, "The visible hand of management replaced the invisible hand of market forces where and when new technology and expanded markets permitted a historically unprecedented high volume and speed of materials through the processes of production and distribution. Modern business enterprise was thus the institutional response to the rapid pace of technological innovation and increasing consumer demand in the United States during the second half of the nineteenth century."[24]

The best solution is probably to say that the relationship between technology and capitalism is, like the relationships between technology, science, and art, interactive. In other words, technology sometimes stimulates capitalism while at other times capitalism stimulates technology. Another way of putting it is that a dialectical relationship exists between the two. In any event I think it's fair to say that the combined forces of technology, science, and capitalism, however you divide them, currently rule the world.

CHAPTER 7

THE NATURE OF TECHNOLOGY

> There is no such thing as "restrained progress."
> You are hearing many voices today that object
> to an "unrestricted technology." A restricted
> technology is a contradiction in terms.
>
> AYN RAND

If technology can be seen as a unified phenomenon, and if there are certain underlying characteristics that all technologies share, then observing those underlying characteristics should allow us to discern patterns that technology consistently manifests over time, especially as it reaches relatively advanced stages of development.

The question I'd like to examine in this chapter is whether those characteristics and patterns suggest that technology possesses a *nature*, just as human beings are said to have a nature.[1]

Jacques Ellul answered that question in the affirmative, as have other philosophers who favor "broad" interpretations of technology. It's their perspective that will be echoed here, with latitude.[2]

What I'm describing here are *general* characteristics of the technological phenomenon *as a whole*. Just as there are exceptions to general descriptions of human nature, and to general descriptions of the nature of men or the nature of women, there will be exceptions to the broad patterns I ascribe to technology. Early in his famous essay "The Hedgehog and the Fox," Isaiah Berlin issued a disclaimer regarding the personality types he identified that applies to the present discussion: "Of course, like all over-simple classifications of this type, the dichotomy becomes, if pressed, artificial, scholastic and ultimately absurd. But if it is not an aid to serious criticism, neither should it be rejected as being merely superficial or frivolous; like all distinctions which embody any degree of truth, it offers a point of view from which to look and compare, a starting-point for genuine investigation."[3]

Four basic, overlapping characteristics or sets of characteristics can be cited as fundamental elements of the nature of technology. They are

1. Technology is by nature expansive.
2. Technology is by nature rational, direct, and aggressive.
3. Technology by its nature combines or converges with other technologies.
4. Technology by its nature strives for control.

To say that technology "strives" for anything is a textbook example of the reifying language that drives people like Joseph Pitt crazy. In answer to those objections, I'll repeat that a systemic interpretation of technology includes the human beings who create and drive technological systems. Such an inclusion suggests that technology as a systemic phenomenon does, in fact, take on the characteristics of a living entity. This, in turn, leads to one of the central questions in the philosophy and history of technology—some scholars consider it *the* central question—and that is whether technology can be said to behave, at a certain stage of development, autonomously. Is it possible, in other words, that at some point technology functions beyond human control?[4]

In this chapter I'll briefly review each of the four basic characteristics of technology's nature. In the next chapter I'll consider the question of technological autonomy.

1. Technology is by nature expansive.

Technique is never content with stasis. It always seeks to widen its sphere of influence. Langdon Winner, a leading philosopher of technology and one of Jacques Ellul's leading defenders, puts it well: "If there is a distinctive path that modern technological change has followed it is that *technology goes where it has never been*. Technological development proceeds steadily from what it has already transformed and used up toward that which is still untouched."[5]

Ellul himself characteristically described this quality in somewhat more dramatic terms. The "proper motion" of technique, he said, is toward "completeness.... To the degree that this completeness is not yet attained, technique is advancing, eliminating every lesser force. And when it has received full satisfaction and accomplished its vocation, it will remain alone on the field."[6]

McDonald's, an enterprise wholly dependent on a massively interconnected complex of technologies and techniques, provides an obvious but nonetheless paradigmatic example. The company will never be satisfied with 100,000 franchise outlets and 20 billion burgers sold (or whatever the current numbers are);

new outlets must constantly be opened in new territories. Although overexpansion and subsequent contraction are possible, in general constant growth is the overriding mission of modern corporate enterprise. Globalization is the natural result of technological expansion and represents the ongoing colonization of the nontechnical world.[7]

The pervasiveness of technique isn't only broad; it's dense. Our day-to-day lives are thick with layer upon layer of technological artifacts and systems, and growing steadily thicker. There seems to be no place in modern society where a TV set or a smartphone doesn't belong, no space advertising can't fill, no patch of ground off limits for development. As technical expansion proceeds, Langdon Winner says, technology becomes "a generalized environment for human existence." Indeed, with global warming we may finally have fulfilled Francis Bacon's stated goal at the outset of the scientific revolution: "the enlarging of the bounds of Human Empire, to the effecting of all things possible."[8]

Nor is technique's tendency to expand limited to physical space. To the contrary, its forward motion includes the occupation of hearts and minds. There's no need to go into this at any length—it's hardly news—but for the moment merely note the never-ending, ever-growing crowd of distractions with which technique competes for our mental attention. A 2012 report by Google warned advertisers that they haven't yet adjusted to the fact that consumers now regularly use as many as four screen devices in their daily lives—mobile phones, tablets, TV sets, and laptop or desktop computers—often more than one of them at the same time. The number of apps available on the Apple iPhone, meanwhile, has grown from 50,000 a year after it was introduced in June 2009 to 1.5 million in June 2015. Note also the ever-growing list of medications prescribed by doctors and therapists to calm us down or pick us up psychologically and emotionally. In these ways and others technique has colonized consciousness as thoroughly as it has colonized nature.[9]

The possibilities for expansion of technology's sphere of influence are endless, not only in computing but also in biology, architecture, transportation, entertainment, education, and a hundred other fields. Tens of thousands of engineers around the world are being paid, day in and day out, to bring those possibilities to fruition. Human progress in general and national economies in particular have become inextricably linked with technological progress, which means we have committed ourselves irrevocably to a future that is even more pervasively technological than the present.[10]

2. Technology is by nature rational, direct, and aggressive.

Individual technologies can be subtle and flexible, but, overall, technique's methods tend to be direct, inflexible, and aggressive—even ruthless. "Technique worships nothing, respects nothing," Ellul said. "It has a single role: to strip off externals, to bring everything to light, and by rational use to transform everything into means. . . . It is rigid in its nature and proceeds directly to its end. It can be accepted or rejected. If it is accepted, subjection to its laws necessarily follows."[11]

Ellul also argued that it is not in technology's nature to make distinctions between moral and amoral use. "Good" for a machine means that it works; "bad" means it doesn't. The single most important value for technique is efficiency—results. Circumstances outside the defined function are irrelevant. A cell phone will ring while you're sitting at your desk or while you're sitting in a funeral. The Internet will just as readily send to your computer a prayer for the day or a video of a woman having sex with a dog.

The quintessential statement of technological aggression may have been uttered by Robert Moses, the master builder of New York City, who said, "When you operate in an overbuilt metropolis, you have to hack your way with a meat ax." Lewis Mumford, following Milton, defined mining as an emblematic technological activity because it uses pick, ax, and shovel to dig deep into the earth and take valuable elements out of it, elements to be forged by fire and anvil into other tools of force and violation. (Fracking and offshore drilling are recent equivalents.) The historian Carolyn Merchant took this imagery a significant step further when she noted that technology is often discussed in language suggestive of rape. Bacon, for example, advised not only that nature must be obeyed, but also that "she" would never give up her secrets willingly. The mechanical arts must be employed, he said, in order to "lay hold of her and capture her" to ensure that she "betrays her secrets more fully." The ambitions of today's digital buccaneers to unleash the next "disruptive" technology are closer to these sorts of impulses than they'd probably care to admit.[12]

Technology absorbs as well as overtakes. It becomes a way of thinking and ultimately a way of being. The German philosopher Martin Heidegger described the technological mind-set as "calculative." It "plans," "investigates," and "computes," he said, and "never stops, never collects itself." He contrasted these characteristics with "meditative" thinking, a mode of consciousness he feared was being permanently eclipsed in the technological era. Meditative thinking "bides its time," waiting to see what will unfold, and in so doing "contemplates the meaning which reigns in everything that is." (Heidegger was notorious for, among other things, his idiosyncratic, often incomprehensible, turns of phrase.) He added that meditative thinking is practical as well as fulfilling, despite the fact that from the standpoint of calculative thinking it tends to be regarded as useless.[13]

Heidegger made a similar distinction between ancient and modern approaches to technological practice. All technology is a form of "revealing" or "bringing into presence," he said. In the classical crafts tradition, an object was allowed to "arise out of itself," much as a flower blossoms. The artist's task in revealing was to assist in the emergence of the object's form from the materials in which it was encased. Modern technique, by contrast, "challenges forth" the materials and reveals by "setting upon." The process is directed at "unlocking and exposing" the energy within the materials—not for its own sake, but for its usefulness somewhere else. Thus nature no longer exists for its intrinsic value, but becomes raw material—"standing reserve"—for exploitation.[14]

A key manifestation of the "setting upon" to which Heidegger referred is the practice of disassembling wholes into component parts in order to facilitate their manipulation. This is the essential logic of the assembly line, of binary computer code, and of technique in general, a fact noted by Adam Smith on the first page of *The Wealth of Nations*: "The invention of all those machines by which labor is so much facilitated and abridged seems to have been originally owing to the division of labor." Marshall McLuhan called this "manufacture by mono-facture," which he defined as "the tackling of all things and operations one-bit-at-a-time." As he put it in 1968, long before digital technologies took center stage, "The breaking up of every kind of experience into uniform units in order to produce faster action and change of form (applied knowledge) has been the secret of Western power over man and nature alike."[15]

The great prophet of manufacture by mono-facture was the American efficiency expert Frederick Winslow Taylor. Taylor was immensely influential in the late nineteenth and early twentieth centuries for applying what he said were scientific techniques to the analysis of the manufacturing process. His techniques weren't rigorous by today's standards, but their detached dissections of individual tasks were substantially more rigorous than the traditions, habits, and intuitions that prevailed before him.

Taylor's method was to divide each factory operation into discrete units to determine, through scrupulous observation and careful measurement, the "one best way" each step could be accomplished. He timed with a stopwatch a workman's every move so that any wasted motion could be eliminated. Any obstacle interfering with the completion of a task was removed; wherever a machine or a procedure could be accelerated, it was. Once the analysis was completed, a set of instructions would be drawn up and posted so that every job could be completed with maximum efficiency, every time. Deviation from the prescribed routine was prohibited.[16]

Taylor embodied the aggression of technological power: everything was sacrificed on the altar of efficiency. When workers balked at the rigidity of his systems, he bought their acquiescence by instituting a graduated pay scale: wages increased when productivity exceeded the required minimum, as defined by Taylor's formulas. This served to increase productivity still further. Henry Ford introduced his famous Five Dollar Day for the same reason: to buy acquiescence to conditions that had been seen as intolerable.

When under attack Taylor regularly claimed that efficiency benefited labor as much as management, but his contempt for the dignity of workers was legendary and often unconcealed. "In our scheme, we do not ask for the initiative of our men," he said in a lecture. "We do not want any initiative. All we want of them is to obey the orders we give them, and do what we say, and do it quick."[17]

The management expert Peter S. Drucker has said that Taylor's methods rank as "the most powerful as well as the most lasting contribution America has made to Western thought since the Federalist Papers." By no means does Taylor deserve full credit for the cult of efficiency that so defines business practice today, however. As Ellul said, efficiency is the lingua franca of the Kingdom of Technique; it naturally emerges as a central focus wherever technological methods are employed. Henry Ford, for example, was obsessed with efficiency as he brought the modern assembly line into being at Highland Park and River Rouge—historian Douglas Brinkley called the drive for efficiency "the essence of Fordism"—but there's no evidence that Taylor's writings influenced him in those pursuits. Ford read little, studied less, and followed his own lead.[18]

Certainly the religion of efficiency continues to exert profound effects on managerial practice among corporate leaders today. The economist Clayton M. Christensen of the Harvard Business School argued in 2013 that American corporations have become so addicted to increasing productivity by employing efficiency methods that they are crippling long-term economic growth.[19]

To say that technique operates by rational means is not to say that it *is* rational. A central theme of romantic objections to the technological project has been the frequency with which so-called reason opens the door to madness. As Norman Mailer put it in his coverage of the Apollo 11 mission to the moon, "A century devoted to the rationality of technique was also a century so irrational as to pin in every mind the real possibility of global destruction."[20]

3. Technology by its nature combines or converges with other technologies.

Technology is irrepressibly catalytic. Invention leads to other inventions. Individual technologies and technological systems can be assembled into new combinations. Techniques invented for one purpose are applicable to any number of other uses and spread outward into the culture, spawning still other uses. These are the dynamics of technological convergence and diffusion, which along with manufacture by mono-facture are the propulsive forces of technological expansion.

An example of technological convergence that's inescapable today is the ongoing blurring of lines between computers, telephones, television, and cable, a combination that has influenced everything from science, news, and entertainment to publishing, education, religion, politics, transportation . . . the list could go on indefinitely. The latest frontier to excite the tech world is the use of computer software to make old-fashioned machines smarter, a campaign that's proceeding under the banner "the Internet of Things." This multitude of permutations has sprung from one seminal technology: the digital micro-processor. The microprocessor may prove to be the most fertile machine of all time, but the intermeshing of techniques and the diffusion of influences it demonstrates can be seen in countless other examples.

An urban airport is a less glamorous manifestation of technological conver-gence. There we find not only airplanes and their associated air-traffic control and maintenance technologies, but road systems and parking lots for cars, rail systems for transport between terminals, vast computer systems for ticketing and baggage, X-ray security systems, public address systems, fast-food outlets, lighting and heating systems, restroom facilities . . . again, the list could be ex-tended indefinitely. All these myriad technologies combined add up to the in-credibly complex entity we call "airport."[21]

Langdon Winner called convergence "the genius of the twentieth century," and many expect it to become an even more dominant force in the twenty-first. Among them are the "transhumanists" who believe that human beings will ul-timately merge with their machines. Ray Kurzweil uses the acronym GNR—for genetics, nanotechnology, and robotics—to describe the three technologies he believes will be most fundamental, together with computing, in bringing the human/machine merger to fruition. The future GNR age, Kurzweil says, "will come about not from the exponential explosion of computation alone but rather from the interplay and myriad synergies that will result from multiple intertwined technological advances."[22]

Technological diffusion describes the spread of a technology from one field of endeavor to another field, often in ways its inventors never imagined. Bronze casting techniques developed for the molding of church bells became useful in the manufacture of cannons, for example. The steam engine was used to pump water out of mines in England for a hundred years before Richard Trevithick attached one to a carriage and put the carriage on a pair of horse-tramway tracks, thereby inventing the railroad.[23]

Technological diffusion radiates outward promiscuously. In addition to its profound influence on computer technology, technologies developed for the American space program have spread into the culture in literally hundreds of products and devices, from sunglasses, mattresses, and running shoes to arti-ficial hearts and global positioning systems. Diffusion can move in more than one direction. A military supercomputer called the Roadrunner has been used to test the viability of the nation's nuclear weapons. It's powered by parallel processing computer chips originally created for the PlayStation 3 video-game console.[24]

Technical ideas diffuse as readily as the techniques themselves. The disassembly lines of Chicago slaughter houses helped inspire Ford's automobile assembly line; McDonald's applied the assembly-line concept to hamburgers. Hundreds of businesses in dozens of fields, from home construction and office supply stores to tax specialists and brake repair shops, have followed the McDonald's franchise model, a technique that has shaped American culture as profoundly as the microprocessor.[25]

Technological convergence often advances out of necessity to correct what the economic historian Nathan Rosenberg has called "technological disequilibrium." Disequilibrium occurs when an innovation in a technology or technique creates a systemic imbalance that has to be adjusted, often by the upgrading of other elements or the addition of new elements in the system. The growing mechanization of the textile industry in eighteenth-century England spurred an ongoing set of innovations, from the fly shuttle and the spinning jenny to the water frame and the cotton gin, to cope with the ever-increasing speed of spinning and weaving and the ever-growing demand for cotton. Slavery was one of the ancillary systems that helped supply that growing demand. The growth of railroads accelerated the growth of telegraph lines because railroad companies needed a signaling system that would help them avert collisions between trains using the same set of tracks. Technological disequilibrium helps explain why there's often a "lag" before a new technology realizes its full potential. Mass production of the automobile required parallel developments of road, fuel, and traffic control systems, all dependent on massive ancillary systems of their own.[26]

The cumulative power of technological convergence and diffusion is demonstrated by the explosion of innovation that followed the emergence of two key developments over the course of the nineteenth century. The first of these was the successful exploitation of standardized and interchangeable mechanical parts. This was an idea conceived in Europe but effectively realized in the United States, beginning soon after the Revolutionary War with the manufacture of firearms. A handful of innovative engineers, usually working separately but often drawing on one another's ideas, designed a series of machines that could be run by unskilled workers—boys, in many cases—to produce individual rifle parts. Those parts could then be easily assembled, also by unskilled workers, into a finished rifle, and the parts from one rifle could be used in any other rifle of the same type. The "American system," as envious Europeans called it, opened the door for the onset of true mass production and helped the new republic establish itself with surprising speed as a world industrial power.[27]

When the concept of interchangeable parts spread to other industries there was an explosion of converging and diffusing technologies based on shared production methods. Nathan Rosenberg has documented direct links from machinery used in rifle manufacture to machinery used in revolvers, locks, sewing machines, bicycles, and eventually automobiles. This diffusion in turn prompted a vast increase in specialization as companies increasingly focused on one part or one machine used by a wide variety of manufacturers. Thus improvements in sewing machines were applied in the production not only of clothing but also of shoes, tents, sails, awnings, saddles, harnesses, handbags, and books. At the same time improvements in the machinery that *made* sewing machines subsequently became useful in machines used to make tools, cutlery, locks, arms, textiles, and locomotives, not to mention other machines. The reign of the skilled craftsman who made entire products by hand had come to an end.[28]

The second key development that led to an explosion of nineteenth-century innovation was the merging of technology with science. In the seventeenth century Francis Bacon had called for the practical application of scientific inquiry—"If it be barren, then let it be set at naught"—but it wasn't until the Industrial Revolution that science and technology were systematically united in common cause. This partnership was especially fruitful in the United States because of the conspicuous pragmatism of its citizens. Benjamin Franklin epitomized the American focus on useful applications of science, a focus noted by Alexis de Tocqueville. "Those who cultivate the sciences amongst a democratic people," he wrote, "are always afraid of losing their way in visionary speculation."[29]

As the century progressed, the synthesis of research and application was driven by advances in chemistry and electricity that made laboratory experimentation as important as mechanical know-how in the development of new products and techniques. The growing sophistication of the machine-tool industry also contributed. If parts were to be truly interchangeable, new levels of precision in measurement and production were necessary, and new, more precise instruments—scientific instruments—were needed to provide them. The practical integration of science and engineering during this period, said the historian Edwin T. Layton Jr., "constituted a change as momentous in its long-term implications as the industrial revolution itself."[30]

A less tangible factor was a new respect for scientific methods overall. Science came to be seen as the new, supposedly objective standard. Exactitude and certainty supplanted the "rule of thumb." Nowhere was this more evident than in the "scientific management" techniques of Frederick Winslow Taylor. "Both sides must recognize as essential the substitution of exact scientific investigation and knowledge for the old individual judgment or opinion, either of the workman or the boss," Taylor told a congressional committee.[31]

An outgrowth of the convergence of science and technology was the transition from independent inventor to corporate research and development, a transition led and embodied by Thomas Edison. There was irony in this because, as mentioned in chapter 2, Edison's public persona was that of the lone genius—"the Wizard of Menlo Park"—when in fact he was instrumental in supplanting the lone genius with the invention factory. Edison's career also epitomized the growing complexity of invention as it expanded from the workshop to massive technological systems. As Edison himself described the challenge of illuminating America's cities,

> It was not only necessary that the lamps should give light and the dynamos generate current, but the lamps must be adapted to the current of the dynamos, and the dynamos must be constructed to give the character of current required by the lamps, and likewise all parts of the system must be constructed with reference to all other parts, since, in one sense, all parts form one machine, and the connections between the parts being electrical instead of mechanical.... The problem that I undertook to solve was, stated generally, the production of the multifarious apparatus, methods, and devices, each adapted for use with every other, and all forming a comprehensive system.[32]

Edison assembled at Menlo Park the engineering and scientific talent and the physical resources he needed to realize such a system. This was a harbinger of the future. Here, too, convergence leading to systemic applications encouraged increasing specialization. Corporate research and development laboratories could afford to assign their salaried scientists to attack specific aspects of an overall project, producing advances that could then be assembled in increasingly complex configurations. As Willis R. Whitney, the founder of one of the first corporate invention factories, General Electric Laboratories, put it, "All new paths both multiply and divide as they proceed."[33]

One particular event stands out for me as a totemic expression of the principle of convergence. In 1908 the Royal Automobile Club of England issued a challenge to the American car company Cadillac. The club's members—in fact, the entire British automotive industry—refused to believe that a car could be successfully manufactured with interchangeable parts. Cadillac was a pioneer in interchangeability. Henry Ford's assembly-line production of Model Ts was still five years in the future, and most cars were still being made essentially by hand. Cadillac executives, eager to prove the superiority of their system, happily accepted the British challenge.

Three Cadillacs were shipped to England and dismantled in a garage near London. The 721 parts of each car were mixed together in a pile on the floor so that no one could tell which part came from which car. Some parts were removed and replaced by stock parts from the Cadillac warehouse to further test the company's claim that off-the-shelf replacement parts would work just as well as the originals. Mechanics then assembled three Cadillacs from the collection of intermingled parts, started them up, and drove them away. The feat won Cadillac the automobile club's prestigious Dewar Trophy and proved once and for all that the American system of mass production was the wave of the future.

Cadillac's expertise in interchangeable parts descended from its founder, Henry Leland, who worked in a factory that mass-produced rifles during the Civil War. Leland applied the expertise learned there to the manufacture of sewing machines, the manufacture of bicycles, and finally to the manufacture of automobiles.[34]

4. Technology by its nature strives for control.

With the exception of the quest for power, to which it's closely related, control is the most fundamental impulse driving technology, starting with our need to control nature. As physically vulnerable creatures, we have relied for millennia on technology as our principal means of survival, and it has served us well.[35]

A survival issue we haven't quite managed to control is the final one, old age, although with the help of plastic surgery, we're doing our best. If the predictions of the transhumanists are accurate, before long we'll be able to avoid dying entirely, either by installing mechanically or biologically engineered replacement parts or by uploading our consciousness into a "platform" more durable than tissue. Thanks to technology, down to our DNA we're being improved: as one enthusiast put it, genetic manipulation will soon allow us to "seize control of our evolutionary future."[36]

In the meantime we're struggling to control our machines. More massive, more powerful, more complex, and more interconnected technological systems demand ever-greater systems of control lest inconsistencies cause equally massive, equally powerful, equally extensive disruptions. Anyone who's sat for an hour in a traffic jam knows how a minor hiccup—a fender bender at rush hour—can bring an entire system to its knees. The classic example is the failure of a relay switch in a power station in Niagara Falls, New York, on November 9, 1965. Within minutes that failure had triggered a series of mechanical and electrical events that blacked out power to the entire northeastern United States and significant portions of Canada.[37]

Webs of interlocking technologies such as those that caused the blackout of 1965 are called "tightly coupled systems" because their various components are so intimately linked that the behavior of any single element affects the other elements in the system. Edison recognized this dynamic and warned against it in his design of the giant machine that comprised his electric lighting system. "Like any other machine the failure of one part to cooperate properly with the other part disorganizes the whole and renders it inoperative for the purpose intended," he said.[38]

The technological society we live in today can be considered one gigantic tightly coupled system. Losing control of any significant part of that system gives rise to a host of frightening scenarios, which is why establishing effective mechanisms of technological control has been a goal of governments and businesses since the Industrial Revolution began. The adoption of standardized time zones as the railroads expanded is an example of the shaping power of those demands.[39]

The desire for control leads to two other characteristic technological drives: the drive toward centralization and the drive toward consolidation. We're told that new technologies open the way for disruption and fragmentation of established industries and institutions, and that's true, but only temporarily; the drive to seize control immediately becomes the drive to retain and extend control. Standardization, economies of scale, and elimination of competition all reduce variety but enhance both stability and efficiency and therefore control. The result is gigantism, as evidenced by the waves of mergers and acquisitions that have periodically characterized industrial and postindustrial capitalism from the days of the robber barons to the present. Among the casualties are independent farmers and the local businesses that in earlier times populated American downtowns. Perhaps the greatest irony of the Internet and its digital siblings has been the concentration of power in the hands of massive multinational corporations that dominate what were supposed to have been the most democratizing technologies in the history of the world.[40]

Testimonies to the glories of the free market are, often as not, at odds with the use of lobbyists, campaign contributions, propaganda, and other techniques aimed at keeping control in the hands of those who already have it. Potential sources of disruption, such as unions and an independent press, are, wherever possible, neutralized. Other common methods of maintaining control include public relations firms and human resources departments. These examples suggest why Jacques Ellul used the term "technique" to describe the broader, not-strictly-mechanical characteristics of technological systems.[41]

Engineers have long recognized that the greatest threat to technical control is human error, which is why the goal in service as well as mechanical systems is typically to minimize as much as possible the exercise of human discretion. As Ellul put it, "The combination of man and technique is a happy one only if man has no responsibility. Otherwise, he is ceaselessly tempted to make unpredictable choices and is susceptible to emotional motivations which invalidate the mathematical precision of the machinery."[42]

Whether or not Jeff Bezos ever read Ellul, those are words he lives by. Amazon is hardly the only company to collect massive amounts of data in an effort to control every movement its employees make, but by all accounts it does so more aggressively than most. A former employee told the *New York Times* that the company "is running a continual performance improvement algorithm on its staff." One can only imagine how fervently Frederick Winslow Taylor would applaud Bezos's application of his ideas.[43]

Where the human being can't be eliminated, he must be adapted. This was the point that aroused Ellul's greatest indignation. All the institutions of society, from schools to religion, have become geared, he believed, to reshaping the human personality to better suit technical demands. "Technique advocates the entire remaking of life and its framework because they have been badly made," he wrote. ". . . The creation of an ideal man will soon be a simple technical operation."[44]

Here, too, Amazon testifies to the accuracy of Ellul's prediction. A plethora of techniques are employed to ensure that workers are pushed to their limit. Those who can't take the pace are quickly terminated; a ruthless evaluation process hastens their departure. Those who can take it become what's known within the company as "Amabots," proud soldiers in Bezos's march to world domination. One former employee described the company's labor relations policy as "Purposeful Darwinism." Survival of the fittest, in others words.[45]

Again, the mistake that is often made is to think that Amazon's use of these sorts of practices is in some way unusual. In fact the company only extends and amplifies, often not by much, techniques and attitudes that are widely manifested not only in corporate life today but throughout the culture as a whole. The influence of technology's character expands along with the expansion of technology itself.

One of the better short statements of that influence I've seen was contained in the presidential address to the Organization of American Historians by John Higham of Johns Hopkins University in 1974. "Even before the Civil War demonstrated that ideology could not by itself permanently hold the country together, another system of integration was emerging," he said.

> The new pattern was one of technical unity. By that I mean a reordering of human relations by rational procedures designed to maximize efficiency. Technical unity connects people by occupational function rather than ideological faith. It rests on specialized knowledge rather than general beliefs. It has had transforming effects in virtually every sphere of life. As a method of production, technical integration materialized early in the factory system. As a structure of authority it has taken the form of bureaucracy. As a system of values, it endorses a certain kind of interdependence, embodied in the image of the machine. Technical relations are machinelike in being impersonal, utilitarian, and functionally interlocking. Since the Civil War the growth of technical unity has been the most important single tendency in American social history; and its end is not yet in sight.[46]

CHAPTER 8

WHO'S IN CHARGE HERE?

The shadow we have wantonly evoked stands terrible
before us, and will not depart at our bidding.

THOMAS CARLYLE

As I mentioned in the previous chapter, the four characteristics I've singled out as constituting the underlying nature of technology collectively point to the central question of whether technology at some point becomes autonomous—that is, does technology at an advanced stage of development become impossible for human beings to control?

Many, if not most, people will find an affirmative answer to that question absurd. Outside of science fiction movies, they'll tell you, machines don't run by themselves. Skeptics might be more receptive if the issue were presented slightly differently: How much *practical* control do we have over the technologies in our lives?

You can choose to live without a car, for example, but not without the pollution cars create. You don't have to watch television, but the politicians who run your government will be elected by people who do. You can choose to live without a computer, but good luck finding a job. Being a committed pacifist won't protect you if the bombs fly. Simply put, our entire social system runs on a vast array of technologies and would collapse within a day or two without them.[1]

I've talked this over with friends who insist they *could* live without cars, or grocery stores, or telephones—most technologies—if they really put their minds to it, and to some extent I think that's true. But by no means would their living standards be in step with even their poorer fellow citizens. Ted Kaczynski was one of the few willing to give it a concerted go, and I had the impression when he was arrested that not a few Americans were almost as horrified by the conditions in which he lived as they were by the crimes he'd committed.

As strange as the idea of technical autonomy might seem, any number of eminently respectable individuals have endorsed it. One was the late historian and head of the Library of Congress Daniel J. Boorstin. In 1977 Boorstin contended that Americans live in a "Republic of Technology." Technology had become the dominant force in our culture, he said, and its proliferation was beyond our control. "We live, and will live," he said, "in a world of increasingly involuntary commitments."[2]

The same opinion had been expressed a decade earlier by the economist John Kenneth Galbraith. "I am led to the conclusion," he said, "which I trust others will find persuasive, that we are becoming the servants in thought, as in action, of the machine we have created to serve us."[3]

The physicist Werner Heisenberg shared Boorstin's and Galbraith's view. In his book *Physics and Philosophy* Heisenberg argued that technological expansion had become a "biological process" that was overtaking even those nations who would have preferred not to pursue it. "Undoubtedly the process has fundamentally changed the conditions of life on earth," he said; "and whether one approves of it or not, whether one calls it progress or danger, one must realize that it has gone far beyond any control through human forces."[4]

None of these opinions would go unchallenged by most philosophers of technology today. The current consensus leans heavily toward the "social constructivist" view, which holds that society shapes technology, not the other way around. Still, the autonomy question remains a controversial one, although usually it's discussed in terms of technological "determinism." That is, to what degree, if any, can we say that technology *determines* the conditions in which it exists?

Autonomy and determinism are not quite the same thing, but they're closely related. Autonomy focuses on the degree to which human beings can or can't control technology. Determinism focuses on the degree to which technology controls human events. Technological determinists believe that technology shapes society. The definitive determinist statement is from Karl Marx, who believed that technical development determines political authority. "The hand-mill gives you society with the feudal lord," he said; "the steam-mill, society with the industrial capitalist."[5]

The social constructivists argue that the influence of a given technology is properly understood only when one understands how it fits into the social, cultural, historical, and physical contexts from which it arose and within which it survives. A given machine or technique may well be one formative influence among many, they believe, but technology never drives history on its own, and technologies are always subject to a variety of influences that help determine when and how they're used.[6]

My own view is that we live in an era of de facto technical autonomy. By that I mean that our ability to choose which technologies influence our lives and how they influence our lives has become, for reasons I'll explain, increasingly limited. We do have choices, no question, but we're not nearly as free to choose as we think. I can restrict my own use of technology, but I have considerably less ability to limit my exposure to the technology that surrounds me. Climate change, depletion of species, the presence of various pollutants in our bodies, and many other environmental repercussions of technology affect literally every person on the planet. That's one aspect of what I mean by de facto technological autonomy. The social and economic demands of the technological society can, theoretically, be avoided, but not without adopting a lifestyle of radical disconnection.

I think it's fair to say that the average American today is a technological determinist without realizing it. By that I mean that most people who scoff at any suggestion we're not in control of our machines will at the same time take it for granted that technology defines us in countless ways. The popular press shares and reflects those assumptions.

"The Computer Age Is Already Changing Every Aspect of Our Lives," declared a 1995 headline in *Newsweek*. "The revolution has only just begun, but it's already starting to overwhelm us," wrote reporter Steven Levy. "It's outstripping our capacity to cope, antiquating our laws, transforming our mores, reshuffling our economy, reordering our priorities, redefining our workplaces, putting our Constitution to the fire, shifting our concept of reality and making us sit for long periods in front of our screens while CD-ROM drives grind out another video clip."[7]

We don't spend time today waiting for our CD-ROM drives to grind out video clips, but that's because we don't need to—the video "clips" are delivered to us by the Internet, the same technology that put *Newsweek* out of business. Meanwhile the latest iteration of the computer revolution promises more of the same. As a 2013 book title put it: *Big Data: A Revolution That Will Transform How We Live, Work and Think.*

Other technologies that are widely assumed to have influenced the direction and character of human affairs include the telephone, the steam engine, the electric light, airplanes, and antibiotics, to name just a few. Abraham Lincoln credited the railroads and the telegraph for helping bond the disparate states of the republic into "one national family"; Marshall McLuhan argued that television was uniting the world into a "global village." The clock has been blamed for implanting in our minds a growing sense of regimentation and abstraction from nature, while the Pill is thought to have changed the sexual behavior and personal relationships of millions of women and men. Other secondary effects said to have been produced directly or indirectly as the result of various technologies include the relocation of millions of workers and their families from farms to cities and the shaping and reshaping of global balances of power depending on which nations have oil resources and/or nuclear weapons and which don't.[8]

I could go on listing these examples indefinitely, and as I say, most people will have no trouble accepting them. Indeed, so seductive is the appeal of technological determinism that it's embraced with equal enthusiasm by those who think technology will save us and by those who think it will destroy us. Perhaps the most fundamental expression of technological determinism is the belief that what set humans apart from the rest of the animal kingdom eons ago—the talent that allowed us to emerge triumphantly from the Darwinian scrum—was our ability to make and use tools.[9]

Not being an academic, I find the antagonisms between the determinist and the social constructivist schools somewhat overheated. Not even the hardest of the so-called hard determinists—and Jacques Ellul's critics would definitely include him in that category—believe that technology literally takes on a life of its own or that it independently determines the course of history. Ellul argued quite clearly that societies are not passive recipients of technology. Rather, he listed a number of conditions that can influence whether technological development is more or less likely to flourish. Chief among these conditions is the degree of "plasticity" a technology encounters in a given social setting, meaning the presence or lack of presence of established institutions, habits of production, religious traditions, trade organizations, and other inhibiting factors that might cause technological change to be resisted. Indeed, historians believe that America's high degree of social plasticity as the Industrial Revolution took off—the relative fluidity of its institutions, the relative freedom of action of its citizens, and the relative powerlessness of its indigenous populations—is a key reason the United States emerged as an international industrial power as quickly as it did.[10]

A related social influence cited by Ellul as triggering receptivity or resistance to technological change is the "technical intention" of a population, meaning its general eagerness to participate in the technological project. Alexis de Tocqueville, Max Weber, and others have theorized that the seeds of the Industrial Revolution were planted by the tenets of Western Christianity, which guided believers to value order, frugality, good works, and human dominion over nature. Whether or not that's true, our encounters with Islamic fundamentalism have proved that degrees of technical intention still vary between and within nations, as they do from region to region, from city to city, and from person to person. If an individual's technical intention is low in a country where the general level of technical intention is high, she'll be dismissed as a "slow adapter" or a "Luddite." Often slow adapters suffer only in becoming the butt of good-natured jokes, but the consequences can be severe. Ask the Native Americans.[11]

The question of whether a given technology exerts an independent, determinative influence on the culture—the question of technological autonomy—comes into play once that technology has become established. There are several reasons why hard determinists such as myself believe that many technologies can and do exert such an influence.

The most basic of these reasons derives from what I call the *facticity* of technology, by which I mean the existence of the tool or the machine itself. If I have a knife or a power saw I am able to effect changes in the world that someone who doesn't have a knife or power saw can't. This ability is deterministic: the possession of the tool can determine an outcome. Similarly, possession of an airplane enables a constellation of activities that would not be possible without an airplane. Langdon Winner points out that in a sense the fundamental purpose of any technology is to have a deterministic effect in the world. We want to accomplish something, and our tools and techniques are means to that end. Obviously the possession of more-powerful technologies can determine more extensive outcomes, although as their impacts broaden, more diffuse causes and effects exert their respective influences. But the facticity of the technological world we've constructed also broadens and remains in place as our new reality.[12]

A key reason established technologies become autonomous is that we come to depend on them, in the process erecting a complex set of supportive technologies that deepen our commitment. Winner calls this the "technological imperative." Most advanced technologies, he says, require the existence of other technologies in order to function as intended. The automobile, as mentioned, requires roads, supplies of gasoline, traffic control systems, and the like. Electrical lighting requires a vast network of power lines, generators, distributors, light bulbs, and lamps, together with production, distribution, and administrative systems to put all those elements (profitably) into place. A "chain of reciprocal dependency" is established, Winner says, that requires "not only the means but also *the entire set of means to the means*."[13]

A social commitment to such a system constitutes de facto autonomy because to undo that commitment would inflict unacceptable damage on the culture. A recent example is the difficulty Japan has experienced giving up its dependence on nuclear power, despite the Fukushima meltdowns that left whole regions of the country uninhabitable. Winner points out that we often become committed to a technology before we realize we're becoming committed to it and also before we fully appreciate the full range of consequences that commitment entails. He calls this "technological drift." The industrial nations of the world committed themselves to technologies of oil and coal before fully appreciating the fact that the greenhouse gases they produce would fundamentally alter the climate of the planet. Now they find it impossible to give up those technologies and the technologies they support because, if they did, their economies would crumble.[14]

The technological imperative is closely related to another manifestation of technical autonomy, "technological momentum." This is the tendency of techniques or machines to become increasingly difficult to displace because of the economic and psychological investments of those who build them. Here the focus is more on the resistance to change by individuals and institutions than on the dependency of those who use the technologies those individuals and institutions provide, although the results of technological momentum affect them both. The historian of technology Thomas Hughes coined the term, and although it describes a key dynamic in the architecture of technological autonomy, it's always struck me as a bit confusing. Momentum in my mind connotes movement, yet in some ways what Hughes is talking about seems more closely related to inertia. Computer programmers refer to the acquired intractability of older software systems as problems of "legacy" or "lock-in," which may be more fitting terms for describing the phenomena in question.[15]

Companies or industries committed to a certain way of doing business are often reluctant to change what they're used to. In this sense the old methods do take on momentum in the way that Hughes suggests, because those habits move forward of their own accord and become increasingly difficult to dislodge or divert. This resistance is manifested materially in the investment of resources in existing plants and procedures and psychologically in the attachments of employees who have invested their careers in those plants and procedures. A classic example is Western Union's passing up the chance to buy Alexander Graham Bell's patent for the telephone for $100,000 in 1876. The company's executives were focused on developing their existing technology, the telegraph, and saw no practical value in Bell's method of transmitting voices, a device they alternately described as "a scientific curiosity" and a "toy." A more recent example is the failure of Xerox to capitalize on innovations in personal computing—the mouse, word processing, and onscreen graphics displays among them—that had been developed by programmers in its own research center during the 1970s. Steve Jobs is said to have been very impressed by all those ideas when he toured the Xerox facility in the formative days of Apple Computer.[16]

For all their collective brilliance, the giants of the computer industry have proved to be singularly susceptible to technological momentum over the years. "Mainframes, microcomputers, PCs, PDAs—at the outset of each innovation, the old guard has fought a pitched battle against the upstarts," notes journalist John Markoff, "only to give in to the brutal realities of cost and performance." This history is undoubtedly a major reason why staying nimble is something of an obsession in Silicon Valley today. Easier said than done, of course. Larry

Page and Sergey Brin, the founders of Google, tried to sell their search algorithm to a number of existing Internet companies, but all passed, saying there was no money in searches. Page and Brin dropped out of Stanford to form a company themselves. They now regularly vow that they're taking steps to ensure that Google will retain, or regain, the speed, flexibility, and daring of a startup.[17]

The processes of convergence and diffusion described in the previous chapter are other powerful engines of technological autonomy. As techniques spread through the culture, a cumulative accretion of applications and effects sets in. Jacques Ellul called this the "self-augmentation" of technique and stressed that it proceeds almost of its own volition: once a technique is invented, it's relatively easy to find another use for it. This, I think, is what technological momentum is really about. In that regard I'd add that while conservative interests often work to keep established technologies in place, a more powerful force is working to overturn them: youth. The forward motion of technological expansion as driven by the enthusiasm of young people for new technologies and for the new opportunities presented by new technologies cannot be overestimated.[18]

It seems obvious that the most reasonable position to take vis-à-vis technological autonomy and determinism is what the historian Edwin Layton calls "reciprocal causality." Society and technology, he says, "are not discrete events but complex combinations of institutions, ideas, values, and events. . . . Certain social forces may lead to technological innovation, which in turn may cause some social changes, and these may in turn produce new technological and social developments. I see no difficulty in the assumption that technology and society mutually influence each other."[19]

Whether technology or society acts as the "primary member" in the relationship, as the philosopher of technology Carl Mitcham puts it, varies in different circumstances, but in every case both are involved. Most people would agree, for example, that the automobile has substantially shaped American society, but most would also agree that a host of nontechnical elements—the country's expansive geography; a willingness to allocate tax dollars to highway construction (and the availability of tax dollars to allocate); the sense of freedom that people gain from driving, and so on—played a codetermining role.[20]

In truth it's impossible to separate social values from technology simply because social values are built into technological devices and systems. The various technologies and techniques that enabled the development of suburbs, for example, simultaneously enabled the realization of two aspirations widely shared by white city-dwellers: the dream of owning a freestanding home with a yard for the kids and the dream of escaping integrated neighborhoods.

Whether the building of the suburbs fueled the desire to escape or merely provided an opportunity to do so, they now exist and have helped define a whole new set of physical as well as social realities.

As I've mentioned, to suggest that our technologies are not in our control strikes some people as an insult to human dignity and an evasion of responsibility. "Guns don't kill people," the saying goes, "people do." I hope what I've said here shows that the question is more complicated than that. The gun and the person who uses it are fused elements in a technological system that creates possibilities that didn't previously exist. Technology enthusiasts often cite the creation of possibility as the most fundamental justification of their enthusiasm, equating possibility with freedom. They forget that possibilities opened can narrow freedoms as well as expand them.[21]

PART III

IN RELATIONSHIP WITH MACHINES

Do not underestimate objects.

DAVID FOSTER WALLACE

CHAPTER 9

QUALITY

> Notice that we're not very much at home
> In the world we've expounded.
>
> RAINER MARIE RILKE

For a time I contributed an occasional essay to a technology blog called *Cyborgology*, which is known among the people who care about such things for its staunch opposition to "digital dualism."

Digital dualists are said to believe that there's an existential difference between being online and being offline. Their assumption is that when we log on we enter another realm, frequently called "cyberspace." It's also assumed that this other realm is a diminished reality relative to the "real" world, a pale facsimile of the world as it exists offline.

The *Cyborgology* crew considers this nonsense. Despite what *The Matrix* and *Neuromancer* tell us, they argue, there's nothing "virtual" about the online experience. It isn't the refuge of lonely souls, huddling in the shadows. We're just as present in the real world when we're posting on Facebook or texting on our smartphones as we are when we're hiking the Grand Canyon.

My own feeling is that the *Cyborgology* folks are correct to a degree. Unfettered involvement in the world *can* include posting on Facebook and Twitter. However, I also think they underestimate the seductive, distancing powers of technology.

The crucial questions in the digital dualism debate involve consciousness and attention. I'm a consciousness conservative in that I believe attention is basically a zero-sum game. Contrary to the multitasking myth, you can't have it both ways: You're either paying attention or you're not. Part of the reason I feel this way is my (fairly extensive) experiences with LSD and my (fairly limited) exposure to Zen, both of which affirm that, even in the best of circumstances, we spend most of our time distracted from fully experiencing what's going on around us.

It's been said many times before but remains true nonetheless: technology promotes distraction. The most fundamental reason this is so is simply that technology relentlessly wants to give us *more*. As any meditation teacher will tell you, *more* is a recipe for distraction.

I've referred already to Robert Pirsig's book *Zen and the Art of Motorcycle Maintenance* and to Pirsig's stated goal in that book of finding a way to resolve the tension between the classic and the romantic points of view. Romantics need to realize, he said, that order and discipline can be a source of beauty as well as power, not to mention lots of useful technological products, including motorcycles and houses that are heated in winter. Classicists need to acknowledge rationalism's limits and its tendency not to recognize its limits.

The vehicle Pirsig prescribed for reconciling the classic/romantic split was something he called "Quality." It wasn't always easy to understand what exactly "Quality" was supposed to mean, in part because for much of the book Pirsig explicitly resisted defining it. My interpretation, one Pirsig would probably find offensively simplistic, is that Quality represents a combination of Caring and Attention. I capitalize those two words to emphasize that Quality will remain elusive unless you *really* care and unless you *really* pay attention. In order to truly pay attention, it's necessary to possess sufficient peace of mind not to be distracted. In order to achieve Quality, then, one must be centered, engaged, flexible, and invested.[1]

When applied to the creation and use of technology, Quality would manifest and exploit the best of both classic and romantic perspectives. Pirsig practiced motorcycle maintenance as a form of meditation in the same way that Steve Jobs tried to instill Apple's products with human values. The title of *Zen and the Art of Motorcycle Maintenance* hinted at that unity, which was also implied in two of the book's most frequently quoted comments: that the Buddha can be found just as easily in the gears of a motorcycle as in the petals of a flower, and that the real cycle you're working on is yourself.

This perspective anticipates and supports *Cyborgology*'s critique of digital dualism. Technology doesn't *necessarily* promote distraction and disengagement. Point taken. Nonetheless, I would argue that Pirsig's meditative approach to motorcycle maintenance reflects an ideal point of view, one that in day-to-day circumstances is difficult if not impossible to maintain. Readers of *Zen and the Art of Motorcycle Maintenance* learn that Pirsig is as vulnerable to the assault of sensory overload as anyone else, if not more so.

The question is why technology so often seems to steer us away from Quality and toward distraction and disengagement. Pirsig tells a story that testifies to the depths of the problem. He'd taken his motorcycle to a repair shop where it was worked on by a team of boy mechanics whose breezy confidence masked a near-total absence of skill or finesse. Three botched overhauls forced Pirsig to return the bike to the shop, where he watched the boy mechanics abuse his machine with casual brutality. He finally fled before further damage could be inflicted.

Reflecting back on the experience, Pirsig decided that a lack of engagement—a Care-less-ness—was the only possible explanation for the mechanics' behavior. The speed with which they worked was one clue to this attitude; the fact that they listened to the radio while they worked was another. But what revealed their disengagement most clearly, Pirsig said, was the expression on their faces, which he described as

> good natured, friendly, easygoing—and uninvolved. They were like spectators. You had the feeling they had just wandered in there themselves and somebody had handed them a wrench. There was no identification with the job. No saying, "I am a mechanic." At 5 p.m. or whenever their eight hours were in, you knew they would cut it off and not have another thought about their work. They were already trying not to have any thoughts about their work *on* the job. . . . They were involved in it but not in such a way as to care.[2]

That story strikes me as both exasperating and sad, mainly because it's so familiar. I'm sure that inattentive, incompetent louts could be found working in the blacksmith shops of medieval villages. Nonetheless, the sort of distraction Pirsig describes seems a peculiarly modern affliction. We regularly encounter facsimiles of those mechanics in retail stores or on the telephone: people who simply don't give a damn about their jobs or the people they're supposed to be helping. It's easy to be annoyed by them and at the same time hard to blame them. We might well act the same way if we were in their shoes. Still, it's fair to wonder: Is there something endemic to the technological society that creates this lack of concern?

There are some obvious and some not-so-obvious answers to that question. As many critics have noted, convenience is an obsession of the consumer culture. The ticket to success in today's marketplace is to sell something that requires no effort or thought whatsoever. The philosopher John Lachs calls this "the fallacy of free delight." Many have also noted two corollaries of the gospel of convenience, especially applicable to mechanical products: ignorance and disposability. Most of us have no idea how the products we use are produced or how they work, and their manufacturers make sure we don't need to know. If they break, they can be replaced rather than repaired. Beneath the fecklessness of convenience there's also an undercurrent of (mostly repressed) fear. On some level we know that ignorance means helplessness. For the vast majority of us there's one option when our computers or our cars or our refrigerators or our toilets or our lights stop working. Call a technician, and hope to hell he comes soon.[3]

Matthew B. Crawford has written an excellent book on this subject, *Shop Class as Soulcraft: An Inquiry into the Value of Work*. In it he points out that, not so long ago, Sears mail-order catalogs included parts diagrams and schematics for the appliances and other mechanical items they carried, on the assumption that if something went wrong, the people who bought them would want to fix them. Appliances no longer come with such diagrams. For Crawford a paradigmatic example of the state of modern disengagement is the fact that some recent cars don't come equipped with a dipstick, thus making it impossible for their owners to perform the rudimentary task of checking the oil.[4]

Our collective disinterest in getting our hands dirty with modern technology marked a crucial turning point in the history of Apple Computer. Apple's earliest computers featured a series of ports that would allow users to adapt the machines as they wished by plugging in peripheral applications. These reflected the hacker sensibility of programming genius Steve Wozniak, who couldn't imagine buying a machine you couldn't *play* with. Steve Jobs's genius was to see that the mass market doesn't care what goes on inside the computer. As he told Walter Isaacson, "My vision was to create the first fully packaged computer. We were no longer aiming for the handful of hobbyists who like to assemble their own computers, who knew how to buy transformers and keyboards. For every one of them there were a thousand people who would want the machine to be ready to be run."[5]

This is lopsidedness from the other direction.

Another important cause of disengagement in the technological society, less often discussed, is retreat. The heads of charity organizations worry about "compassion fatigue," when people get worn down by hearing too many voices pleading for help and eventually shut all of them out. That's a manifestation of the zero-sum game I mentioned: There's only so much caring and attention any of us has to give. On an infinitely broader scale, this is what's happening to all of us in the technological society. Functioning in that society confronts us with an overwhelming flood of demands on our attention. After a while, in any number of ways, we begin to shut down. Caring and attention dissipate. We disengage.

In their book, *The Meaning of Things: Domestic Symbols and the Self*, Mihaly Csikszentmihalyi and Eugene Rochberg-Halton explore the sorts of relationships people develop with objects in their homes. They describe these objects as signs, or "objectified forms of psychic energy." A transaction occurs, they say, between ourselves and the things we possess. We "charge" the objects around us with psychic energy, and those objects return that energy in a sort of ongoing feedback loop.[6]

Csikszentmihalyi and Rochberg-Halton argue that by definition our store of psychic energy is depleted when we invest some of that energy in the objects around us. How much we give of ourselves, and how much we resist giving ourselves, depends on what we think we will get in the exchange. The psychic energy drained by our investment in an object or task may be turned into a net gain if the energy we invest is returned in the form of a sense of accomplishment or satisfaction.

Another way of describing this psychic energy is in terms of meaning. We invest objects with meaning depending on what they represent to our sense of self and to our relationships with others and with the world. Advertising agencies make good use of this proclivity. Status symbols—expensive cars, jewelry, and the like—are one way in which objects can manifest meaning, but they're far from the only way. A wedding ring may carry greater emotional than financial value, for example. Observing the various ways in which people interact with objects helps us understand, Csikszentmihalyi and Rochberg-Halton write, "the process by which we become human."[7]

In all these forms of investment we see the presence of care. Csikszentmihalyi and Rochberg-Halton give the example of a china plate that has been passed down through generations. The plate takes on special meaning not only because it's old and beautiful, but also because of its fragility. The fact that it has survived intact suggests the care that has been invested in it. By contrast, a set of wrenches that has been passed from grandfather to father to son might take on special meaning because of their durability. In both cases objects acquire meaning by being kept and being given. Presents are objects invested with meaning.

Problems occur, however, if an object requires an investment of psychic energy that we give unwillingly, or that is not returned with an equal or greater amount of energy. Nothing can drain energy faster than trying to install a new software program that doesn't work. A thoughtless gift can be dispiriting rather than uplifting. The result, Csikszentmihalyi and Rochberg-Halton say, is "psychic entropy," which can take the form of boredom, alienation, or frustration, all of which may ultimately manifest themselves as disengagement, or a lack of caring. Robert Pirsig describes a series of "gumption traps" that can drain the energy needed to properly focus on motorcycle repair. These include, in addition to setbacks encountered in the job itself, distractions, insecurities, or assumptions that erode the quality of attention we bring to the job.[8]

It's important to emphasize that, while Csikszentmihalyi and Rochberg-Halton's analysis focuses on objects that people choose to have in their homes, exchanges of psychic energy occur with any object we encounter in our environment, outside as well as inside the home. Some of those reactions are conscious but most are unconscious. We spend a tremendous amount of psychic energy in the technological society filtering out sensory input so that we don't have to pay conscious attention to it. That's one reason travel can be so exhausting. We're not sure what we need to pay attention to in the waves of unfamiliar information coming at us, so we have to consciously process more of it than usual. We decorate our homes to be protective cocoons, selecting objects that provide a meaningful buffer against the world.

The comforts and tensions produced by our encounters with objects in different settings underscore the fact that we are *in relationship* with the things around us, a relationship that constitutes a form of ongoing conversation. Csikszentmihalyi and Rochberg-Halton argue that the meaning of objects we encounter is not limited to the meanings we project onto them. Objects make their own, independent contribution to the conversation by projecting into the world their inherent character and characteristics. Objects also arrive with built-in meanings "scripted" into them by the culture. "Without doubt," Csikszentmihalyi and Rochberg-Halton write, "things actively change the content of what we think is our self and thus perform a creative as well as a reflexive function."[9]

The idea that objects project meaning into the world finds support in philosopher Davis Baird's concept of "thing knowledge." Simply by being, Baird says, artifacts communicate knowledge of how they were built and why they work the way they do. Those who know how to "read" this knowledge can understand it as effectively as a printed diagram or a spoken explanation. That is why an inventor, simply by looking at a device, is able to incorporate its elements or principles into another device. It's also why companies can reverse engineer their competitors' products. Baird's idea was anticipated by Henry Ford, who insisted that inventions are far more than lifeless objects. "You can read in every one of them what the man who made them was thinking—what he was aiming at," he said. "A piece of machinery or anything that is made is like a book, if you can read it. It is part of the record of man's spirit." This extends Marshall McLuhan's famous maxim that the medium is the message. The same can be said of any technology.[10]

It's easy to take for granted the abundance the technological society offers (another symptom of lack of attention), but it helps to recognize how radically the quantity of artifacts in our lives has accumulated. Csikszentmihalyi and Rochberg-Halton point out that, except in the higher reaches of the aristocracy, few homes in the Middle Ages had any furniture at all. People generally sat and slept on sacks of grain or mounds of straw. Plates, silverware, clothing, and "any other kind of movable property" were also exceedingly rare, they add. The Industrial Revolution touched off a simultaneous "Consumer Revolution," flooding first Europe and then America with tidal waves of goods that have only grown more overwhelming ever since.[11]

It would be hard not to consider the comforts produced by that revolution a blessing—it's nice having chairs, beds, and dishes. At the same time many of us have come to feel smothered by stuff, not only in our homes but in the sea of machines, sounds, images, and structures in which we're daily immersed.

Harper's Magazine published a superb essay on this subject a few years ago, "The Numbing of the American Mind: Culture as Anesthetic," by Thomas de Zengotita. It opened with an epigraph from Friedrich Nietzsche: "The massive influx of impressions is so great; surprising, barbaric, and violent things press so overpoweringly—'balled up into hideous clumps'—in the youthful soul; that it can save itself only by taking recourse in premeditated stupidity."[12]

By taking recourse in premeditated stupidity, de Zengotita explained, Nietzsche meant that we allow ourselves to become ignorant by becoming anesthetized, a condition Nietzsche attributed to the fact that, as de Zengotita put it, "people at the end of the nineteenth century were suffocating in a vast goo of meaningless stimulation." It goes without saying that the goo has grown infinitely thicker since then. Marshall McLuhan made an observation similar to Nietzsche's more than half a century later, regarding the sensory overload poured on us by radio and television: "We have to numb our central nervous system when it is extended and exposed [via communications technologies], or we will die. Thus the age of anxiety and of electric media is also the age of the unconscious and of apathy."[13]

For many centuries monks have withdrawn from civilization in order to focus on union with God. The Industrial Revolution gave birth to a more secular brand of asceticism, one aimed at preserving sensibility by avoiding the corrosive effects of overstimulation. A pioneer of this method was John Ruskin, who despised riding on the railroad precisely because it jammed too many sensations into too short a time. "If the attention is awake, and the feelings in proper train," he said,

> a turn on a country road, with a cottage beside it, which we have not seen before, is as much as we need for refreshment; if we hurry past it, and take two cottages at a time, it is already too much; hence to any person who has all his senses about him, a quiet walk along not more than ten or twelve miles of road a day, is the most amusing of all travelling; and all travelling becomes dull in exact proportion to its rapidity.[14]

Ruskin was one of the leading figures of the Arts and Crafts movement, which aimed to counter the drift into disengagement and the consequent erosion of quality with hands-on attention to craftsmanship. Those values were revived by the counterculture of the 1960s. Thousands of youths eschewed established professional pursuits, not to mention the military, to become potters, painters, bakers, weavers, jewelry makers, and leatherworkers. Some of them made a go of it; many others eventually turned to more conventional jobs, meaning jobs more directly related to the functioning of the technological society.

One person who managed to avoid that path is Wendell Berry, who since 1965 has worked a farm near Port Royal, Kentucky. "My own experience has shown that it is possible to live in and attentively study the same small place decade after decade, and find that it ceaselessly evades and exceeds comprehension," Berry has written. ". . . A place, apart from our now always possible destruction of it, is inexhaustible. It cannot be altogether known, seen, understood, or appreciated."[15]

It's common to blame disengagement these days on digital devices, and without question their ubiquity and their mobility have vastly increased the density of the technological tendrils with which we've entangled ourselves. But as I've repeatedly argued, digital technologies aren't the only technologies that matter, and we were beginning to show the strain long before the microprocessor entered the picture. More than a century before Robert Putnam published *Bowling Alone*, for example, Ralph Waldo Emerson described detachment as the predominate mood of his age. "Instead of the social existence which all shared was now separation," he wrote. And long before digital dualism became an issue, Max Frisch described technology as "the knack of so arranging the world that we don't have to experience it."[16]

No doubt city life seemed bewildering to country bumpkins long before the Industrial Revolution (Lewis Mumford cites an Egyptian papyrus that contains "a stinging indictment" of the miseries of urban life), but it's fair to assume that urbanity's capacity for bewilderment increased as modernity progressed. Certainly the number of bumpkins who experienced it did. At the beginning of the twentieth century the philosopher Georg Simmel identified a "blasé attitude" as a defining characteristic of city dwellers trying to cope with the overstimulation and anonymity of metropolitan life. A new sort of perception resulted, he said, from an "*intensification of nervous stimulation* which results from the swift and uninterrupted change of outer and inner stimuli."[17]

Obviously it was the tender-minded, to use William James's phrase, who found it hardest to hold aloft the shield of nonchalance. Here is William Wordsworth's description of living in London:

> Oh, blank confusion!
> true epitome
> Of what the mighty city is herself,
> To thousands upon thousands of her sons,
> Living amid the same perpetual whirl
> Of trivial objects, melted and reduced

> To one identity, by differences
> That have no law, no meaning, and no end[18]

Matthew Arnold wrote of the same weariness:

> For what wears out the life of mortal men?
> 'Tis that from change to change their being rolls;
> 'Tis that repeated shocks, again, again,
> Exhaust the energy of strongest souls,
> And numb the elastic powers.[19]

As Arnold's poem suggests, overstimulation isn't the only reason we disengage. Dislocation does it, too. Change has always been a constant of human affairs, but as we all know, technology has amplified the pace and scale of change exponentially. It's interesting that Alvin Toffler's concept of "future shock" doesn't get talked about much anymore, despite the fact that the acceleration of technological change responsible for that state of psychic imbalance has, as he predicted, only increased in the decades since he coined the phrase.

No one knows this better than those inside the maelstrom. Before he stepped down as CEO of Sony Corporation in 2012, Howard Stringer told the journalist Ken Auletta that he'd all but given up trying to stay abreast of every new gadget to hit the market, even though he headed one of the leading technology companies in the world. "There are so many options out there, simultaneously, that it's a dizzying experience," he said.

> For every time you see an opportunity, you also see a threat. Every time you see a threat, you see an opportunity. That's the one-two punch of the technological marathon we're all in. You worry about missing a trend. You worry about not spotting a trend. You worry about a trend passing you by. You worry about a trend taking you into a cul-de-sac. It means that any CEO or senior executives of a company have to induce themselves to have a calm they don't feel, in order to be rational in the face of this onslaught.[20]

Stringer had inherited what has become a tradition of disinheritance. Compare his sentiments with a comment made in 1890 by the Cornell University economist Jeremiah Jenks: "No sooner has the capitalist fairly adopted an improved machine than it must be thrown away for a still later and better invention which must be purchased at dear cost if the manufacturer would not see himself eclipsed by his rival."[21]

A common literary theme that emerged in parallel with the Industrial Revolution concerned the anxiety and loss people felt as the world they thought they knew disappeared, seemingly overnight. The poet laureate of displacement is Henry Adams, who in the opening pages of his autobiography, published in

1907, described himself wondering from an early age, "What could become of such a child of the seventeenth and eighteenth centuries, when he should wake up to find himself required to play the game of the twentieth? . . . No such accident had ever happened before in human experience. For him, alone, the old universe was thrown into the ash-heap and a new one created."[22]

Adams was far from alone, but it was no surprise he felt that way. A sense of isolation was and is a key symptom of the modern condition. The nineteenth-century version of "future shock" was "anomie," a term coined by the French sociologist Émile Durkheim and defined by Merriam-Webster as "social instability resulting from a breakdown of standards and values" as well as "personal unrest, alienation, and uncertainty that comes from a lack of purpose or ideals." In 1897 Durkheim published a study in which he blamed an alarming increase in suicides across Europe on the "morbid disturbance" caused by "the brilliant development of sciences, the arts and industry of which we are the witnesses." The work of centuries, he said, "cannot be remade in a few years."[23]

We know now that there would be little attention devoted to remaking the work of centuries or to repairing the morbid disturbances that accompanied their demise. Observe, for example, the strikingly similar sentiments expressed by the historian Arthur Schlesinger Jr. on the occasion of America's Fourth of July celebration in 1970 (the same year *Future Shock* was published). The country was experiencing an "extreme crisis of confidence," he said, one provoked by the "incessant and irreversible increase in the rate of social change" caused by technological advances that "make, dissolve, rebuild and enlarge our environment every week."[24]

The moneyed classes, of course, have always had means to shield themselves from the discomforts of disruption. Those of the original Gilded Age retreated into what Lewis Mumford called "a cult of antiquarianism," celebrating medieval chivalry and piety while lounging in Victorian drawing rooms that excluded, in Mumford's words, "every hint of the machine." Even there, a new affliction, "neurasthenia," invaded. The term was popularized in a best-selling book titled *American Nervousness*, published in 1884. Its author was the physician George A. Beard, who was a friend of Thomas Edison's. Beard described neurasthenia as an overload of the body's electrical network, "the cry of the system struggling with its environment." Symptoms included "lack of nerve force," insomnia, fear of responsibility, fear of being alone, fear of being in society, indecisiveness, and hopelessness.[25]

Even the era's leading apostle of progress, Herbert Spencer, expressed his concern about the toll that rapid industrialization was taking on the national psyche. This was partly because he'd grown alarmed at the number of nervous breakdowns and suicides he was witnessing among American businessmen and partly because Spencer himself suffered from nervous exhaustion. "Everywhere I have been struck with the number of faces which told in strong lines of the burdens that had to be borne," Spencer said in a speech in 1882. Perhaps, he added, the gospel of hard work had been taken far enough. "It is time to preach the gospel of relaxation." His concerns were widely shared. As a writer for *Harper's* put it in 1894, "Something must be done—this is universally admitted—to lessen the strain in modern life."[26]

Again, abundant evidence testifies to the fact that the strain of modern life has, if anything, only increased. Evidence of that can be gleaned from our pervasive consumption of antidepressant drugs and from the statistics that document sharp rises in the rates of depression and suicide over the past several decades. Nor has technology delivered on its promise to free us from lives of drudgery. Polls find that a majority of Americans hate their jobs or have mentally "checked out" of them. Critics, meanwhile, note that tastes in popular culture have taken a distinct turn toward nostalgia. The past, unlike the present, offers something to hold on to.[27]

Robert Pirsig and Matthew Crawford are far from the only ones to propose conscious engagement as the antidote to semiconscious disengagement. In *Being and Time*, Martin Heidegger took issue with the Cartesian formula, "I think, therefore I am," arguing that "I am, therefore I think" was closer to the mark. In other words, the ability to think abstractly derives from practical consciousness, not the other way around. This bears some resemblance to Hegel's belief that a lack of facility with the physical world inevitably leads to a reversal in the master-slave relationship, the master becoming increasingly disengaged from reality and thus increasingly dependent on the slave, who remains engaged. Another philosopher who recognized the shaping impact of engagement was Karl Marx. "The mode of production of material life conditions the social, political and intellectual life process in general," he wrote. "It is not the consciousness of men that determines their being, but, on the contrary, their social being that determines their consciousness."[28]

In *Technology and the Character of Contemporary Life*, Albert Borgmann provides a remarkable series of excerpts from the writings of a British wheelwright named George Sturt, who described how the character of his livelihood changed as the pace of industrialism accelerated in the early years of the twentieth century. Sturt wrote of his "affection and reverence" for the "tree-clad country-side" that surrounded his village and for the people who lived there, describing his searches for timber through "sunny woodland solitudes," "winter woods," "leafless hedgerows," and "wet water-meadows." The timber he harvested, he emphasized, was neither his "prey" nor a "helpless victim to a machine." Rather Sturt saw the wood as a partner, willing to "lend its subtle virtues to the man who knew how to humor it: with him, as with an understanding friend, it would cooperate."[29]

The tone turned somber when Sturt described the changes that occurred in the wake of World War I, when increased mass-market production meant increased demand for raw materials, including wood. He wrote angrily of a "greedy prostitution" that "desecrated" his beloved forests, of "fair timber callously felled at the wrong time of year, cut too soon." In the shop, work once done by hand was taken over by machines, which drove through timber "with relentless unintelligence," increasing output at the expense of engagement.

"Of course wages are higher," Sturt said. "Many a workman to-day receives a larger income than I was able to get as 'profit' when I was an employer. But no higher wage, no income, will buy for men that satisfaction which of old—until machinery made drudges of them—stream into their muscles all day long from close contact with iron, timber, clay, wind and wave."

Virtually everything Sturt said could easily be put in the mouth of a twenty-first-century environmentalist and just as easily dismissed as hopeless "tree hugger" clichés. Complaints about digital distraction are often dismissed as clichés, too. But something becomes a cliché because it's deemed so commonplace as to be obvious. That doesn't mean it's false.

Csikszentmihalyi and Rochberg-Halton's phrase "psychic entropy" speaks to the limits many of us face in our working lives today, when the prospect of maintaining anything close to a relationship of Quality with the materials at hand is dim, at best. It's one thing to focus intently and caringly on the motorcycle you're riding on vacation, or on the wooden desk you're building in your workshop at home, or on the rows of tomatoes and beans you're growing in your garden. It's another to focus caringly on an assembly-line job in which you perform one isolated task on a product that passes by on a conveyor belt a thousand times a shift or to maintain a sense of reverent intimacy with the textures of a telephone headset and a keyboard as you listen to the disembodied complaints of the fiftieth customer you've talked to over the past several hours.

Here's a quote from the first person profiled in Studs Terkel's 1972 book of interviews, *Working*, a man named Mike LeFevre who spent his days hauling piles of metal from place to place in a steel mill. "You can't take pride anymore," he said.

> You remember when a guy could point to a house he built, how many logs he stacked. He built it and he was proud of it. . . . It's hard to take pride in a bridge you're never gonna cross, in a door you're never gonna open. You're mass-producing things and you never see the end result of it. I worked for a trucker one time. And I got this tiny satisfaction when I loaded a truck. At least I could see the truck depart loaded. In a steel mill, forget it. You don't see where nothing goes.
>
> I got chewed out by my foreman once. He said, "Mike, you're a good worker

but you have a bad attitude." My attitude is that I don't get excited about my job. I do my work but I don't say whoopee-doo. The day I get excited about my job is the day I go to a head shrinker.[30]

We see much the same anomie in Terkel's interview with Beryl Simpson, who worked for twelve years as a telephone representative for an airline. During her time there the company installed Sabre, one of the first computerized reservations systems. "Sabre was so expensive, everything was geared to it," Simpson said.

> Sabre's down, Sabre's up, Sabre's this and that. Everything was Sabre. With Sabre being so valuable, you were allowed no more than three minutes on the telephone. You had twenty seconds, busy-out time it was called, to put the information into Sabre. Then you have to be available for another phone call. It was almost like a production line. We adjusted to the machine. The casualness, the informality that had been there previously was no longer there. . . .
>
> They monitored you and listened to your conversations. If you were a minute late for work, it went into your file. I had a horrible attendance record—ten letters in my file for lateness, a total of ten minutes. . . . When I was with the airlines, I was taking eight tranquilizers a day. . . . I had no free will. I was just part of that stupid computer.[31]

Here again the sentiments expressed have become so commonplace as to be clichés and thus easily dismissed. Nonetheless, they describe the experience of millions of workers today. In an article for the *New York Times* in 2014, labor consultants Tony Schwartz and Christine Porath cited numerous polls showing widespread feelings of disengagement and dissatisfaction among workers, including executives. "The way we're working isn't working," they wrote. "Even if you're lucky enough to have a job, you're probably not very excited to get to the office in the morning, you don't feel much appreciated while you're there, you find it difficult to get your most important work accomplished, amid all the distractions, and you don't believe that what you're doing makes much of a difference anyway. By the time you get home, you're pretty much running on empty, and yet still answering emails until you fall asleep." A follow-up column by Schwartz a week later quoted several of the hundreds of "overwhelmingly acid" comments he'd received from readers in response to the first column, describing their own bitter on-the-job experiences. Schwartz described the response as "a collective howl of powerlessness, despair, cynicism and rage," and added, "I spend all of my working days thinking about this issue. Even so, I was stunned to discover how pervasively and deeply so many people feel victimized, diminished and disempowered by the work that consumes the biggest portion of their waking hours."[32]

The disengagement of the worker from the work is not an accident. Rather it is an intentional and concerted strategy, developed in the industrial era and since applied to the "knowledge" and service workers of the postindustrial "information society." The scientific management techniques of Frederick Winslow Taylor demonstrated an approach to labor that had been identified by Adam Smith and Karl Marx: efficiency of production can best be realized by splitting manufacture into a series of isolated, specialized functions, and the worker must be fitted to the demands of the machine rather than vice versa. Both principles divided work from thought, much less caring, and thus were recipes for disengagement. The technological economy has flourished on that policy and continues to follow it faithfully.[33]

While Steve Jobs recognized that consumers today aren't interested in looking inside their computers, he also knew that on some intuitive level they respond to products that are designed and manufactured attentively and carefully. Quality, in Pirsig's sense of the word, is what made Apple's products stand out in the Jobs era. Many other companies follow that ethos, but many do not. Quantity rather than quality is the defining characteristic of the technological society as a whole. We respond with our own lack of caring and attention.[34]

There's a shallowness to this sort of existence that plays out not only in our relationships to machines but also in our relationships to each other. The theologian Paul Tillich recognized the trajectory. "The decisive element in the predicament of Western man in our period," he said, "is his loss of the dimension of depth."[35]

CHAPTER 10

ABSORPTION

Those who make them are like them;
so are all who trust in them.

PSALM 115:8

So, if attention is a zero-sum game, and if the attention we devote to technology tends to distract or disengage us from the world at large—from what could be described in technological terms as the wide-screen version of existence—what I'd like to talk about here is the other side of the equation: absorption.

To say that we've wrapped ourselves in machines—at home, at work, and everywhere in between—is at this point the most common of commonplaces. Nonetheless, it's worth standing back for a moment to marvel at how quickly and how completely we've been overtaken. Smartphones and iPads, text and tweet, earbuds and Androids, Blu-ray and Bluetooth, Xbox and On Demand. How are we tethered? Let me count the ways.

Digital devices didn't invent technological absorption, but they've made it infinitely more portable and more ubiquitous. The question is, how did it happen? Is absorption an inherent quality of technology? Do our devices demand that we surrender consciousness—large chunks of it, anyway—at the door?

A good place to begin our inquiry is with the emergence of the original computer hackers. They were the pioneers. They threw themselves into the digital maw before most of us knew there was one. They didn't invent technological absorption, but by attaching it to computers they increased its power exponentially. In the process they opened a path that most of us have followed.

Steven Levy's *Hackers: Heroes of the Computer Revolution* documents the evolution of hacker culture from its origins in the late 1950s through the early 1980s. As the subtitle suggests, it's a sympathetic portrait. Nonetheless, Levy's description of the hacker personality conforms to the usual stereotypes of the prototypical nerd. Hackers, he said:

- Are socially inept outside the circle of hackers, which is exclusively male.
- Exist primarily on junk food.
- Pay scant attention to personal hygiene.
- Have a compulsion for figuring out mechanical and especially electrical systems of all kinds.
- Play games obsessively.
- Are fond of elaborate pranks that demonstrate technical skill.
- Are arrogant in the extreme when it comes to proficiency at programming, which is all they really care about.[1]

Levy added that many of the hackers he interviewed had, since childhood, grown accustomed to building science projects while the rest of their classmates were "banging their heads together and learning social skills on the fields of sport." Accustomed to being outcasts, they found in programming a source of tremendous personal power.

I've mentioned that the computer freaks of the counterculture saw computers as tools for expanding consciousness and promoting social change. That's not what the hackers profiled by Levy were into. The hackers' goal in programming was to successfully program, period. Creating programs that would have some utility outside of the computer didn't interest them. One hacker spent all night writing a program that would change Arabic numbers to Roman numerals, an accomplishment that had no purpose other than to prove the hacker could make the computer do it.[2]

For the early hackers getting time on a computer was a major challenge. Personal computers were a distant dream in the early 1960s. University mainframes were locked away in air-conditioned rooms, to be used only by a privileged few. There were time slots, sign-up sheets, and hierarchies, and hackers usually ranked low in the pecking order. So precious was the programming time they were allotted that they devoted intense concentration not only to the hour they were actually seated at the console but also to the hours they spent preparing for their session and to the hours they spent analyzing it once it had ended.[3]

Through their dedication to what they called "the Hands-On Imperative" the hackers achieved an intimacy with their machines, but the relationship was a more disembodied variety of intimacy than that experienced by George Sturt in the woods. "When you programmed a computer," Levy wrote, "you had to be aware of where all the thousands of bits of information were going from one instruction to the next, and be able to predict—and exploit—the effect of all that movement. When you had all that information glued to your cerebral being, it was almost as if your own mind had merged into the environment of the computer."[4]

Levy added that this state of consciousness persisted after the programming session ended—it was as if, having mentally merged with the machine, the hackers left their brains with it even after they'd gone home. The "logical mind-frame required for programming spilled over into more commonplace activities," Levy said. "You could ask a hacker a question and sense his mental accumulator processing bits until he came up with a precise answer to the question you asked."[5]

These tendencies have been described in somewhat more dramatic fashion by Joseph Weizenbaum, a computer sciences professor at MIT. Weizenbaum was a programmer from an earlier generation who began to notice the odd ways in which some students were interacting with computers. In a famous passage from his 1975 book, *Computer Power and Human Reason*, Weizenbaum characterized the emerging breed of computer geek as

> bright young men of disheveled appearance, often with sunken glowing eyes . . . sitting at computer consoles, their arms tensed and waiting to fire their fingers, already poised to strike at the buttons and keys on which their attention seems to be as riveted as a gambler's on the rolling dice. When not so transfixed they often sit at tables strewn with computer printouts over which they pore like possessed students of a cabalistic text. They work until they nearly drop, twenty, thirty hours at a time. Their food, if they arrange it, is brought to them: coffee, Cokes, sandwiches. If possible they sleep on cots near the computer. But only for a few hours—then back to the console or their printouts. Their rumpled clothes, their unwashed and unshaven faces, and their uncombed hair all testify that they are oblivious to their bodies and to the world in which they move. They exist, at least when so engaged, only through and for the computers.[6]

Steven Levy says this passage deeply offended some of the hackers at MIT, who felt it referred directly to them. Weizenbaum denied it and said the description was based on his own experience as a programmer as well as on his observations of others, inside and outside MIT. Whatever the case, there seems little doubt that computer programming can be and often is an obsessive enterprise, a fact that hackers typically consider a point of pride as much as embarrassment. Here, for example, is how Bill Gates described the process by which he and his partner, Paul Allen, wrote the original software for the first personal computer, the Altair 8800:

> Writing good software requires a lot of concentration, and writing BASIC for the Altair was exhausting. Sometimes I rock back and forth or pace when I'm thinking, because it helps me focus on a single idea and exclude distractions. I did a

lot of rocking and pacing in my dorm room the winter of 1975. Paul and I didn't sleep much and lost track of night and day. When I did fall asleep, it was often at my desk or on the floor. Some days I didn't eat or see anyone. But after five weeks, our BASIC was written—and the world's first microcomputer software company was born. In time we named it "Microsoft."[7]

Reading comments such as these, one can't help but think that the original computer hackers were right in considering themselves, as Levy put it, "the vanguard of a daring symbiosis between man and machine." Whether that symbiosis was something to be celebrated is another question.[8]

There's no doubt that focusing intently on the task at hand can be a wonderful thing. Paying attention produces not only great technologies but also great art, great science, great sports, great sex ... great lots of things. Attention plus Caring equals Quality. There's also no doubt, however, that if taken too far, focused attention can lead to a narrowing of perspective that impoverishes other areas of life. Focus becomes absorption.

My concern, shared with many others, is that the promiscuous spread of technology in general and digital technologies in particular has created a state of affairs in which the sort of narrowing we saw in the hackers has become, rather than an on-again, off-again experience of a fairly limited subset of the population, a chronic condition for the culture as a whole.

Granted, the attention to devices we see around us every day is a watered-down version of hacker obsession. It's more like fixation—not much Quality is involved. Still, a form of absorption is at work there—again, not just individually but in the culture as a whole—if only in the quantity and quality of attention *not* being paid to anything else.

Note Steven Levy's comment, in the passage quoted above, regarding the frame of mind the hackers adopted in order to successfully synchronize their thinking with the computer: "Inevitably that frame of mind spilled over to what random shards of existence the hackers had outside of computing." That, in a nutshell, is the problem.[9]

Absorption was identified as an aberration by at least one skeptic from the earliest days of the modern era. Jonathan Swift's *Gulliver's Travels*, first published in 1726, includes an extended satire of Baconian and Newtonian science in general and of the proceedings of the Royal Society in particular. On his third voyage Gulliver is marooned by pirates, then rescued by the inhabitants of Laputa, an island that floats above the Earth. Laputa is ruled by a class of men who are so lost in "intense speculations," one eye turned permanently inward, the other toward the skies, that they are completely unaware of what's happening around them. In order that they might be alerted when someone is trying to speak to them or when they're in danger of walking off a cliff, these gentlemen are followed by "flappers" who stand ready to gain their attention by tapping them on the side of the head with pebble-filled bladders. The frustrated wives of Laputa are delighted to meet strangers, Swift explains, "for the Husband is always so rapt in Speculation, that the Mistress and Lover may proceed to the greatest Familiarities before his Face, if he be but provided with Paper and Implements, and without his Flapper at his side."[10]

Swift's satire was aimed at the craze for experimentation inspired by Francis Bacon's scientific method, but the tunnel vision he described—the mind-set Heidegger would later call "calculative"—was also, as Bacon urged, aimed at "useful" invention. And indeed, productive absorption has been the engine of technological and economic advance ever since.

A passage in Mark Twain's *Life on the Mississippi* provides a classic example of productive absorption at work. It begins with Twain's description of how, during the course of his apprenticeship as a riverboat pilot, he learned to read the river like a book, recognizing in every ripple of its surface warnings of hidden dangers or reassurances of safe passage. Passengers were illiterate in this language, he said, and saw only "pretty pictures ... painted by the sun and shaded by the clouds."[11]

Twain was proud of the expertise he'd gained, but he recognized it had come at a price: "I had lost something which could never be restored to me while I lived. All the grace, the beauty, the poetry had gone out of the majestic river!"[12]

He goes on to describe from memory a glorious sunset he'd watched when he was still a novice: the water at dusk turning from blood red to gold, the lines radiating delicately outward from the black shadow of a log floating on the surface, the glimmers of silver on the leaves of trees lining the riverbank. "I stood like one bewitched," he recalled.

> I drank it in, in a speechless rapture. The world was new to me, and I had never seen anything like this at home. But as I have said, a day came when I began to cease from noting the glories and the charms which the moon and the sun and the twilight wrought upon the river's face; another day came when I ceased altogether to note them.[13]

In *The Machine in the Garden*, Leo Marx comments that the tension Twain described between naive and trained experience was an "omnipresent" theme in nineteenth-century culture, a theme that reflected "a conflict at the center of American experience."[14]

Charles Lindberg showed that these tensions carried over intact into the twentieth century. In his memoir, *The Spirit of St. Louis*, Lindbergh described in the present tense what he saw and felt as he looked down from the cockpit during his first airplane ride: "Trees become bushes; barns, toys; cows turn into rabbits as we climb. I lose all conscious connection with the past. I live only in the moment in this strange, unmoral space, crowded with beauty, pierced with danger."[15]

In the five years of flying experience between that moment and his solo flight across the Atlantic, Lindbergh recognized that his perspective had irrevocably changed.

> I was a novice then. But the novice has the poet's eye. He sees and feels where the expert's senses have been calloused by experience. I have found that contact tends to dull appreciation, and that in the detail of the familiar one loses awareness of the strange. First impressions have a clarity of line and color which experience may forget and not regain.
>
> Now, to me, cows are no longer rabbits; house and barn, no longer toys. Altitude has become a calculated distance, instead of empty space through which to fall. I look down a mile on some farmer's dwelling much as I would view that same dwelling a horizontal mile away. I can read the contour of a hillside that to the beginner's eye looks flat. I can translate the secret textures and the shadings of the ground. Tricks of wind and storm and mountains are to me an open book. But I have never realized air or aircraft, never seen the earth below so clearly, as in those early days of flight.[16]

My favorite example of the triumph of the technological gaze comes from a pilot not of riverboats or prop planes but of space capsules. In *The Right Stuff*, Tom Wolfe describes astronaut John Glenn's reactions to the view outside his window as he orbited Earth—the first American and only the third human to do so—in 1962. Glenn was under specific instructions from his superiors to describe the sights and sounds of his journey so that the taxpayers at home could share in the excitement—and assure funding for additional missions. Enthusing over scenery wasn't something test pilots were inclined to do; such sentimentality violated what Wolfe called the Fighter Jock Code. Glenn, however, was more open than other pilots might have been to public relations initiatives and so was prepared to give it a shot, even though a long checklist of tests and readings also demanded his attention.

Unfortunately, Glenn didn't feel very inspired by the spectacle of the planet unfolding below him. The reason, he eventually realized, was that he'd seen the same view too many times, in satellite photographs and flight simulators. "He knew what it was going to look like," Wolfe (with his trademark stylistic flourishes) said. "It had all been flashed on the screens for him. . . . *Yes . . . that's the way they said it would look. . .* Awe seemed to be demanded, but how could he express awe honestly? He had lived it all before the event. How could he explain that to anybody? The view wasn't the main thing, in any case. The main thing . . . *was the checklist!*"[17]

In fairness I should note that the experience of seeing Earth from space has been powerful enough to make other astronauts look up from their control panels. Some say they've experienced life-changing shifts in perspective upon seeing Earth as a "tiny blue dot" floating in the vast reaches of the cosmos. This "overview effect" reportedly causes them to appreciate the interconnectedness of life on the planet and the absurdity of international conflict.[18]

The dichotomy of these views testifies to the polarity that's embodied in our technologies. As the philosopher of technology Don Ihde has pointed out, when a device amplifies human experience in one way, it inevitably reduces it in other ways. Looking through a microscope allows us to see microorganisms, but while looking at them we can no longer see the table we're sitting at or the room we're sitting in. Ihde concluded that technology's powers of amplification tend to be noticed while its reductions tend to be ignored.[19]

In *The Homeless Mind: Modernization and Consciousness*, Peter Berger, Brigitte Berger, and Hansfried Kellner argue that individual consciousness in the technological society becomes segmented into "clusters" of knowledge related to specific functions, much as the functions in a factory are segmented. The separation between work knowledge and home knowledge (known in the nineteenth century as the doctrine of "separate spheres") is the most obvious and most important of these clusters. It's considered desirable to segregate these functions efficiently; the healthy worker, we're told, will leave her personal problems at home and her work problems at the office. Those who succeed at this segregation have effectively "engineered" their identities, the authors say. Technology has now tipped that balance toward the "always on" side of the equation; the promulgation of digital devices has blurred the work/home distinction by allowing us to be on duty wherever we are. Many Silicon Valley companies, meanwhile, have endeavored to close the circle entirely by creating office spaces that are so comfortable and so much fun that their employees won't *want* to go home.[20]

Long before the digital revolution, Berger, Berger, and Kellner pointed out that strict boundaries between functional aspects of the self can be difficult to maintain. The operative form of knowledge, they say, is often "carried over" from one arena to another and tends to push other types of knowledge into the background. What's carried over isn't specific items of knowledge but the general "cognitive style" that pertains to a given type of knowledge. Thus a person who's accustomed to being a problem solver at work tends to be a problem solver at home.

Charles Lindbergh fit that pattern. It's well known that Lindbergh was an early devotee of the safety checklists that would become part of every pilot's (and every astronaut's) flight routine. Less well known is the fact that Lindbergh followed the checklist philosophy as diligently with his family as he did with his airplanes. His youngest daughter, Reeve, writes in her memoir that soon after his return from a trip, Lindbergh would ritually summon each of his five children, one by one, into his office. There the checklist under each of their names would be systematically reviewed. Entries ranged from specific repri- mands ("chewing gum") to full-scale lectures ("Freedom and Responsibility"). Some items would be checked off, meaning they'd already been addressed, but most would not be. The longer the list, the longer the session in Daddy's office and the more work to be done afterward. Thus for Reeve Lindbergh, seeing her father coming home became less cause for celebration than "an invitation to anxiety, a degree of tedium, and a certain measure of gloom."[21]

I noted earlier that a need for control is a characteristic of technology's na- ture. Not infrequently absorption is applauded as a means of maintaining con- trol and thereby achieving success. Not infrequently we also see absorption spilling over into obsession. These are the hallmarks, and often the downfall, of lopsided men.

Melville's characterization of Captain Ahab provides the classic literary ex- ample. Ahab is consumed by his quest for the whale; nothing else matters. So powerful is this obsession that Ahab succeeds in enlisting the willing coopera- tion of his crew in what only a few recognize as an insane and fatal pursuit. At one point he compares the forward motion of his will to that of a locomotive. "The path to my fixed purpose is laid with iron rails," he says, "whereon my soul is grooved to run."[22]

Ahab notes, too, with a mixture of triumph and surprise, how readily all the crew members of the *Pequod* have acquiesced to his demands. "I thought to find one stubborn, at the least," he says, "but my one cogged circle fits into all their various wheels, and they revolve."[23]

CHAPTER 11

DREAMWORLD

Who is prepared to take arms against
a sea of amusements?

NEIL POSTMAN

Ray Kurzweil's name has come up several times already in this book, usually in fairly dismissive terms. For skeptics like me he makes an easy target because his enthusiasms are so unrestrained and because he's been such a tireless self-promoter. Anyone with a scintilla of scientific credibility who promises that human beings will soon become immortal super-beings can count on more than a scintilla of media exposure.

Kurzweil takes plenty of flack from various experts who say his expectations for advances in their fields are unlikely to impossible, but many others, fellow worshipers in what Jaron Lanier calls the church of "cybernetic totalism," take him seriously. The latter group includes the founders of Google, who have hired Kurzweil to enhance their search techniques with the latest developments in artificial intelligence.[1]

My own objections to Kurzweil have little to do with whether his predictions will come true—I've read enough history to know that betting against techno-logical advance is a fool's game. What bothers me is his almost complete disin-terest in contemplating the infinite ramifications that will unfold—some fore-seeable, some not—if his predictions *do* come true. His gleeful anticipation of the day when virtual reality will overtake everyday reality is an example.

The virtual reality future Kurzweil envisions bothers me because I feel as if American culture is pretty thoroughly saturated with fantasy, falsity, sensation, titillation, and escapism already. That there might be a limit is, of course, not a possibility the technology enthusiast will find the least bit persuasive.

What exactly are the virtual delights Kurzweil believes we'll soon be enjoying? He says that after the Singularity there will be "no distinction" between physical and virtual reality. Nanobots implanted in our brains will be able to suppress signals transmitted by our senses and replace them with synthetic sensations engineered to produce virtual realities. Our brains won't know the difference. Kurzweil calls this "full immersion" virtual reality and says the range of synthetic realities we can experience will be virtually (no pun intended) unlimited. We'll be able to assume any number of different personalities and project different images of those personalities to other people sharing our virtual reality, simultaneously. "Your parents may see you as one person," he says, "while your girlfriend will experience you as another."[2]

This, I hear you saying, is already the case, but wait: it gets more complicated. Your parents and your girlfriend will be able to "override" the personality you're projecting, if they choose to, in favor of one they prefer, which means that in any given interaction a multiplicity of personalities, some "real" and others imagined, could be "interacting" with one another at the same time. It doesn't seem to have occurred to Kurzweil how ripe such a scenario would be for confusion and misdirection, as if relationships aren't complicated enough as it is.

No matter, the development of virtual reality technologies is progressing at a rapid pace, as evidenced by Facebook's $2 billion acquisition of industry leader Oculus VR in 2014. In announcing the purchase, Facebook CEO Mark Zuckerberg sounded positively giddy with excitement at virtual reality's transformative potential. "One day, we believe this kind of immersive, augmented reality will become a part of daily life for billions of people," he said. The expectations of Oculus's chief executive, Brendan Iribe, sounded no less ambitious. "In virtual reality, you are going to find yourself reminding your brain that this is not real," he said. "It's a paradigm change."[3]

Technologists talk about changing paradigms like most of us talk about changing socks, but in the case of virtual reality, I suspect the hype is justified, for one basic reason: it will satisfy a deep-seated hunger that human beings have harbored for millennia. We're eager to surrender our grasp of the actual. If an avenue of escape is offered, we'll take it. Virtual reality is a uniquely effective escape mechanism because it mounts an especially intimate, all-encompassing assault on what are supposed to serve as our anchors to the world-as-it-is: our senses. And it turns out our senses aren't nearly as reliable as we like to think.

An encounter with a primitive virtual reality technology more than a decade ago taught me how easily our senses can be fooled. I'd taken my son, who was seven or eight years old at the time, to visit the Empire State Building in New York City. Upon arriving we discovered that you had to wait for what looked like hours to buy a ticket for the observation tower. However, you could skip the line if you paid something like five bucks apiece extra and sat through a "virtual tour" of the city called the "New York Skyride." Being averse to lines, I forked over the extra money.

My son and I were soon ushered into a small, darkened theater where a series of about a dozen nondescript benches faced a screen that was, I'd guess, about twenty feet high. The setting was more carnival fun house than high tech. It turned out that didn't matter.

The virtual tour recreated a helicopter ride over Manhattan during which the pilot swooped through the city's fabled concrete canyons. At appropriate moments, the seats shook, tilted, or heaved in sync with the point-of-view visuals onscreen. Those simple cues created an amazingly visceral, lifelike feeling. It wasn't total immersion reality, but it was close enough that when the helicopter made a sharp turn or a quick plunge, my stomach did a flip-flop and an involuntary "whoa" escaped from my mouth. I doubt the film was more than ten minutes long, but it was a hell of a lot of fun while it lasted.

I learned that day how little trust we can put in our senses. If equipment that cheap can provide illusions that effective, I don't doubt that the sorts of full-immersion virtual realities Ray Kurzweil talks about will be every bit as convincing as he says they'll be. I'm certain, too, that hundreds of software engineers are working diligently, right now, to devise the software and hardware to make those illusions possible.

An article titled "At the Heart of It All: The Concept of Presence" provides an unapologetic description of the goals those engineers have been aiming for. The paper was written by Matthew Lombard and Theresa Ditton of Temple University and published in 1997 in the *Journal of Computer-Mediated Communication*. (Yes, I think the fact that there is such a journal tells us something.) Their first sentence set the tone: "A number of emerging technologies including virtual reality, simulation rides, video conferencing, home theater, and high definition television are designed to provide media users with an illusion that a mediated experience is not mediated, a perception defined here as presence." The authors subsequently expand on this definition: "An 'illusion of nonmediation' occurs when a person fails to perceive or acknowledge the existence of a medium in his/her communication environment and responds as he/she would if the medium were not there."[4]

According to Lombard and Ditton, there are two ways in which this illusion of nonmediation can occur: When the illusion becomes so transparent that the person forgets that he or she is engaged with a medium, or when the medium appears to be transformed into something other than a medium. The effectiveness of the illusion will increase to the degree that "sensory breadth" and "sensory depth" are achieved. Sensory breadth is defined as "the number of sensory dimensions simultaneously presented"; sensory depth is "the resolution within each of these perceptual channels." Visual and auditory stimuli are the most important, of course, but "olfactory output, body movement (vection), tactile stimuli, and force feedback" can also contribute, as long as each is presented in appropriate balance. The authors conclude that for an illusion of presence to be maintained, "the medium should never draw attention to itself or remind the user that she/he is having a mediated experience."[5]

Put it all together and you have a fully articulated range of techniques of sensory deception. Some might argue that these aren't all that different than techniques employed in the theater for thousands of years to encourage the suspension of disbelief. To that I would answer with the Hegelian maxim that at some point a change of quantity becomes a change in quality. The virtual reality techniques being developed today and promised for the not-too-distant future will be far more powerful than anything the stage can offer. They will also be infinitely more accessible. If Ray Kurzweil is correct, the day is coming when even our remote controls will be unnecessary.

As I say, we humans seem to have an inborn appetite for the suspension of disbelief. For centuries we've sought respite from reality in various forms of ecstatic and narcotic experience. The reasons seem simple enough: life is difficult, and death awaits. So it is that those seeking to create virtual realities have the deck stacked in their favor. We *want* to believe what they're selling.

Anne Foerst is a self-described "robotics theologian" who worked in the Artificial Intelligence Laboratory at MIT during a period when designers there were making robots they hoped would come across as warm and friendly to humans, rather than threatening. In her book *God in the Machine: What Robots Teach Us about Humanity and God*, Foerst described her surprise at discovering how easy it was to respond emotionally to the robots. She felt happy when one of them "smiled" at her and disappointed when one of them "ignored" her, despite knowing that whatever the robots did was programmed behavior that had nothing to do with her personally. The psychologist Sherry Turkle and others have described the same experience.[6]

I was struck by one of the more existential reasons Foerst offered to explain this odd emotional connection. "As a deeply lonely species," she said, "we have a strong desire to communicate with beings different from us." This is a manifestation, she believes, of the same drive we have to connect with our pets, as well as with whales, dolphins, and other wild animals. Foerst also believes that the ease with which we can be deluded in our interactions with machines may be related to the fact that human beings are by nature storytellers. She and others contend that creating realities through metaphor and narrative are among our most basic tools for making sense of the world.[7]

A well-known paper by the psychoanalyst Jacob Arlow takes this a step further, suggesting that in our thoughts we're constantly crossing the line between fantasy and reality. There's an ongoing interplay between the "outer eye" of our sense perceptions and the "inner eye" of our imaginations, Arlow said. Fantasy "exerts an unending influence on how reality is perceived and responded to." Freud explained our ongoing defiance of the "logic of facts" as a coping mechanism in the face of fear, specifically fear of "the superior power of nature, the feebleness of our bodies, and the inadequacy of the regulations which adjust the mutual relationships of human beings in the family, the state and society." Thus we see again that those seeking to create virtual realities have the deck stacked in their favor. We're naturally inclined to buy what they're selling.[8]

This same eagerness to suspend disbelief manifests itself in the emotional attachments television viewers develop to the characters in their favorite programs. We'd like to think this is a habit limited to couch-bound shut-ins of marginal intelligence, but a series of studies by two researchers at Stanford University, Byron Reeves and Clifford Nass, suggest it's far more pervasive. The title of the book they wrote about their research pretty much tells the story: *The Media Equation: How People Treat Computers, Television and New Media like Real People and Places.*

Reeves and Nass put their test subjects into a wide variety of situations in which they interacted in various ways with computers and other technologies. Their goal was to see if people observed the same rules of social behavior with machines as they do with other people. They did. For example, test subjects consistently reported feeling more favorably disposed toward computer programs that flattered them about their performance on a test than they did toward computer programs that didn't flatter them. Similarly, test subjects consistently had more positive feelings about programs that were polite than they did about programs that were neutral. Other experiments showed that test subjects were more likely to be critical of a computer program's performance if their evaluations of that performance were registered on a different computer or on a written evaluation. If the evaluations of a computer were completed on the same computer that was being evaluated, they were consistently more favorable.

Reeves and Nass concluded from these experiments that confusion between mediated life and real life is the rule rather than the exception. The media equation referred to in their book's title is media = real life. This willingness to treat machines as sentient beings didn't depend on the technical sophistication of the media being used, nor did responses vary between people of different ages, educational levels, or professions, including technology experts. "*All* people automatically and unconsciously respond socially and naturally to media," Reeves and Nass said, adding that these responses are "fundamentally human."[9]

The reason that's so, they suggest, is that we respond to modern media with ancient instincts. The human brain evolved in a world in which all perceived objects were real physical objects. Thus we automatically tend to accept that what *seems* real *is* real. It's possible to think your way around this automatic response, the authors say, but to do so takes conscious and sustained effort. In general, they believe, people don't make that effort. Automatic assumptions go unnoticed and unchallenged. Accepting fabrication as reality is the "default" mode.

One of the earliest and best-known examples of how easily our default mode can deceive us is the ELIZA computer program, designed in 1966 by Joseph Weizenbaum of MIT. ELIZA was supposed to come across to users as an automated psychotherapist, although Weizenbaum's real interest was to see how successfully a computer can converse with a human being. The program was named after George Bernard Shaw's character Eliza Doolittle, who, as Weizenbaum put it, could be taught to speak but not necessarily understand. Test subjects would type in comments about what was bothering them; ELIZA would pick up on key words in what they'd written and respond with questions or answers generated by a preprogrammed script. If the "patient" said he was depressed, for example, ELIZA might ask, "Why do you tell me you are depressed?" If the patient typed in, "Everyone is laughing at me," ELIZA might ask, "Who laughed at you recently?"

Weizenbaum assumed that the limitations of the program's responses would quickly become obvious to the people who used it, but that isn't what happened. People responded so positively to ELIZA that when they learned Weizenbaum was reading transcripts of their exchanges with the program, he was accused of violating the bounds of doctor-patient confidentiality. "I knew of course that people form all sorts of emotional bonds to machines," Weizenbaum later wrote, "for example, to musical instruments, motorcycles, and cars. And I knew from long experience that the strong emotional ties many programmers have to their computers are often formed after only short exposures to their machines. What I had not realized is that extremely short exposures to a relatively simple computer program could induce powerful delusional thinking in quite normal people."[10]

Combining the research of Reeves, Nass, and Weizenbaum with the prophecies of Kurzweil should concern anyone who believes that voters ought to be thinking clearly on election day. In 1899 John Dewey wrote that democracy requires "tools for getting at truth in detail, and day by day, as we go along. . . . Without such possession, it is only the courage of the fool that would undertake the venture to which democracy has committed itself." By 2004 such sentiments were deemed obsolete, at least according to an official in the (successful) reelection campaign of President George W. Bush. "Judicious study of discernible reality" is "not the way the world really works anymore," he said, adding that in the modern media environment, "we create our own reality."[11]

This affirmed yet again the prescience of Jacques Ellul, who in 1954 wrote that distortion of the news represents the first step toward "a sham universe," a step that would lead inexorably to "the disappearance of reality in a world of hallucinations." Thirty years after that, in *Amusing Ourselves to Death: Public Discourse in the Age of Show Business*, Neil Postman observed that "disinformation" had become the stock in trade of television news. "Disinformation does not mean false information," he said. "It means misleading information—misplaced, irrelevant, fragmented, or superficial information—information that creates the illusion of knowing something but which in fact leads one away from knowing.... Ignorance is always correctable. But what shall we do if we take ignorance to be knowledge?"[12]

Postman argued that Aldous Huxley's dystopian vision in *Brave New World* anticipated the direction in which the culture was headed far more accurately than George Orwell's dystopian vision in *1984*. "What Orwell feared were those who would ban books," he said. "What Huxley feared was that there would be no reason to ban a book, for there would be no one who wanted to read one. Orwell feared those who would deprive us of information. Huxley feared those who would give us so much that we would be reduced to passivity and egoism. Orwell feared that the truth would be concealed from us. Huxley feared that the truth would be drowned in a sea of irrelevance."[13]

Postman added another observation of Huxley's, from *Brave New World Revisited*: that those who would oppose tyranny tend to overlook "man's almost infinite appetite for distractions." It seems obvious, then, that the more you multiply the sources of distraction, the less capable we become of coherent response to situations that would demand attention if we *weren't* distracted. As Postman put it, "Orwell feared that what we hate will ruin us. Huxley feared that what we love will ruin us."[14]

While modern media's advances in the arts of distraction are new, the appetite they fulfill is not. In fact, it's not going too far to say that the projection of dreams is an inherent quality of the technological project. Philosophers have long argued that a fundamental distinction between science and technology is that science aims at discovering what is while technology aims at providing the means to construct the world according to our desires. Food, shelter, and safety from predators were the first priorities, but a host of other desires followed soon after, including the desire to placate the gods and the desire to escape the quotidian. The pyramids of Egypt were aimed at creating awe, I'm sure, as were the great cathedrals of Europe. These were our prototype virtual-reality environments.

A visitor to the Great Minster Cathedral in Zurich on the eve of the Protestant Reformation would have experienced state-of-the-art virtual reality technology, circa 1500. The interior was almost literally covered with images. Elaborate crucifixes or busts of holy figures ornamented every altar, and there were seventeen altars. Pillars were decorated with scenes from scripture and portraits of saints; murals depicting more scenes from scripture and more saints covered the twelve choir arcades. Most elaborate of all was the Chapel of the Twelve Apostles. Its entryway was covered with murals of Christ with his disciples. At the chapel's center was the "Easter Grave," a sunken chamber accessible only by a sharply descending set of stairs, the sides of which were adorned with paintings of the risen Lord. Wooden statues of Mary, Mary Magdalene, and St. John surrounded the tomb. In it, wrapped in a white coverlet with silken tassels, lay a replica of the body of Christ. On Easter Sunday this replica "rose" from the grave and was paraded in triumph through the church.[15]

This was the sort of sensory extravaganza that caused the newly empowered Protestants to launch their campaign to remove icons from the churches, a reprise of the iconoclastic campaigns waged in Byzantium seven hundred years earlier. Ulrich Zwingli engineered the passage of reforms that prohibited the use of religious imagery anywhere in Zurich, and soon the walls of the Great Minster were painted white. The effect, Zwingli said, was "positively luminous."[16]

The problem with icons, the Protestants believed, is that people had a tendency to worship the image, rather than what the image was supposed to represent. Muslims and Jews have avoided that danger by prohibiting images of sacred figures. Orthodox Christians, by contrast, believe icons serve as windows through which the sacred can gaze upon the faithful, and vice versa. This explains the flatness with which the Orthodox deities and saints are depicted. Their purpose is not to draw viewers into a virtual reality but rather to provide a transparency through which the Holy Spirit can pass.[17]

The Industrial Revolution introduced a new religion to challenge the old: consumerism. As the ranks of its proselytizers and converts grew, techniques of virtual reality, steadily enhanced by the increasingly sophisticated technologies of both manufacture and presentation, were applied in its service.

In *Dream Worlds: Mass Consumption in Late Nineteenth-Century France*, Rosalind H. Williams describes the heightened level of marketing expertise that appeared in the Paris expositions of 1899 and 1900. Whereas expositions of previous years had focused on the latest developments in science and technology, she says, visitors in those years found themselves surrounded by "a fantasy world of pleasure, comfort and amusement" where

> reveries were passed off as reality. They saw crowds milling around displays of luxurious automobiles and around glass cages displaying couturier-clothed mannequins; taking imaginary voyages via cinematic techniques to the floor of the sea or the craters of the moon; and, at night, staring at displays of lighted fountains or at voluptuous belly dancers wriggling in a reproduction of a Cairo

nightspot. The expositions and similar environments (such as department stores and automobile trade shows) displayed a novel and crucial juxtaposition of imagination and merchandise, of dreams and commerce, of collective consciousness and economic fact.[18]

The alliance of capitalism and technology would prove endlessly prolific in the distribution of virtual realities, especially as mass media and advertising came into their own. International expositions turned into world's fairs and then theme parks. In the history of all-encompassing virtual environments, Disneyland was a landmark; Las Vegas may be the (pre-Singularity) apotheosis.[19]

A common characteristic of virtual realities is that their ability to transfix tends to deteriorate, so that ever more elaborate and powerful techniques must be employed to achieve the same impact. Indeed, we can see in the evolution of everything from architecture and video games to fast food and religion how the purveyors of experience have progressively increased the volume of their delivery. "Concern with *effect* rather than *meaning* is a basic change of our electric time," said Marshall McLuhan.[20]

Another common characteristic of virtual realities is that they allow us to avoid confronting life's more unpleasant contingencies. As a result virtuality becomes an increasingly popular refuge when reality becomes especially challenging. This was a tendency noted by Sherry Turkle in her 1995 study, *Life on the Screen*. Much of the book focuses on Multiple User Domains, or MUDs, which at the time served as a principal online gathering place for large communities of fantasy game players. Participants created, through text descriptions typed into their computers, personae that interacted with the personae of other players. The acronym originally stood for Multi-User Dungeons because the first MUDs were dedicated to the game Dungeons and Dragons.

Because players created their own characters and responded in real time to other players, MUDs were a more active form of virtual reality than movies and more dependent on players' imaginations. Later, more sophisticated video games would make it easier to escape, but Turkle's research showed that the fantasy worlds of the MUDs were already engrossing enough to get lost in. Players were known to spend days at a time online, ignoring such distractions as meals and school, and to feel far more fulfilled in their fantasy identities—individual players often had several—than they did in real life. One player used the common MUD abbreviation for real life when he told Turkle, "RL is just one more window, and it's not usually my best one."[21]

If game players were aware that their commitment to the MUDs might stunt their development elsewhere, it was a trade-off they seemed willing to make. Turkle quotes an online discussion in which habitués of a MUD debated how constructive it was to spend so much time "in" (not on) the computer. "Well," said one participant, "hasty judging people might say that the escapists are weak and can't stand the reality—the truly wise see also the other side of the coin: there must be something wrong with Reality, if so many people want to escape from it."[22]

Once again, identical feelings were often expressed as the Industrial Revolution began to manifest the full measure of its power in the late nineteenth century. I mentioned earlier the "cult of antiquarianism" that emerged in the Gilded Age: those who could afford to spent their evenings in overstuffed Victorian drawing rooms, hidden behind brocade curtains that shut out all traces of the upheaval outside—a virtual reality, you might say. Also prevalent during this period, says the historian T. J. Jackson Lears, was a widespread longing for the "dreamlike irrationality" of the Middle Ages; Wagner's mythic-heroic operas and James M. Barrie's *Peter Pan* were exemplars of the theme. Lears quotes Bryan Hooker, an author of medieval romances, who in 1908 remarked on the "curiously vivid" interest of his contemporaries in fairy tales and adventure stories. "Perhaps our very materialism is responsible for this new hunger after fantasy," Hooker observed. "Because the world, never so bluntly actual as now, is too much with us, we spend our vacations upon the foam of perilous seas."[23]

The technology of the motion picture arrived at just this propitious juncture to provide what was, up to that point, our most effective ride into the realm of the fantastic. The popularity of movies was immediate and immense: the number of American movie houses rose from five thousand in 1907 to eighteen thousand in 1914. Perhaps the main reason for this success, according to Neil Gabler, author of *Life: The Movie*, was film's unprecedented power to create virtual realities. Flickering images in dark theaters, Gabler said, "cast a spell that lulled one from his own reality into theirs until the two merged." From the film industry's earliest days this hypnotic effect was a source of concern among social critics. Jane Addams, founder of Chicago's Hull House, called the movie theater a "veritable house of dreams" that was "infinitely more real than the noisy streets and crowded factories."[24]

The attraction persists to this day, as we've progressed from silent pictures to sound and from black and white to color and now 3D. Our appetite for perilous seas seems to have persisted as well, as a glance at the attractions currently playing at your local multiplex will attest.

What sorts of virtual realities will we soon be plugging directly into our brains? According to Ray Kurzweil, the possibilities are endless. Meanwhile the world can go on without us. A comment in Donna Haraway's famous essay, "The Cyborg Manifesto," comes to mind. "Our machines are disturbingly lively," she said, "and we ourselves frighteningly inert."[25]

CHAPTER 12

ABSTRACTION

> The intoxicated soul wills to fly above space and
> Time. An ineffable longing tempts him to indefinable
> horizons. Man would free himself from the earth,
> rise into the infinite, leave the bonds of the body, and
> circle in the universe of space amongst the stars.
>
> OSWALD SPENGLER

If it's true that technology serves as a vehicle through which we project our imaginations, it's also true that a fundamental property of that projection—action at a distance—leads progressively over time to ever-greater levels of abstraction.

We can trace this progression back to the beginning of the relationship between humans and their tools. It was more effective for our earliest ancestors to throw a stone or shoot an arrow at a mastodon than to try to kill it with their bare hands. The primary advantage those technologies offered was force, but action at a distance was also key. Arrow and stone were safer ways—more convenient ways, you could say—to achieve the desired result.

Jump forward to the present day, when we prefer to get our meat shrink-wrapped at the supermarket and to eliminate our enemies with ballistic missiles or, more recently, with drones. This is what I mean when I say the technological project has been characterized by a progressive process of disengagement via action at a distance. The view we get of people walking on the street as we look down from a skyscraper illustrates the basic principle. Detail and connection get lost.

Also lost, often, is awareness of our connection to the results of actions we've initiated through technological agency. The importance of this loss of personal connection—in its impact not only on our personal consciousness but on our relationships to other people and to the rest of the world—cannot be overemphasized. In his 1981 book, *Responsibility and the Individual in Modern Society*, John Lachs identified mediation as a central issue of our time. "Our psychic distance from our deeds renders us ignorant of the conditions of our existence and the outcome of our acts," he said. "... As consciousness of the context drops out, the actions become unmeaning motions without consequence. With the consequences out of view, people can be parties to the most abhorrent acts without ever raising the question of their own role or responsibility."[1]

Obviously, the arrow and stone examples represent a nascent level of interpolation between human being and nature. I'm not arguing that a hoe or a shovel impose a significant degree of disengagement between a farmer and a row of beans. Nor am I saying that action at a distance is necessarily bad. I'm glad I don't have to kill my meat with my bare hands. I'm only noting the path we embarked on. Abstraction increased as our tools drew on sources of power other than our bodies. Thanks to digital technologies, effective action at a distance has progressed to the point today where abstraction has become one of the driving forces of technological progress and therefore one of the driving forces of the culture.[2]

Two related digital technologies in particular are currently at the forefront of this movement: Big Data and computerized simulation techniques. Big Data refers to the use of computer algorithms to find patterns in massive amounts of information. With literally billions of digitized searches, messages, and transactions transmitted daily by millions of individuals worldwide, an incomprehensible amount of information is available today to collect, digest, and exploit. Simulation programs construct computer models that can provide pictures of how some object or process—the design of a building, the strategy of a military invasion, the absorption of a drug by the digestive system, the movement of a weather system, votes in an election campaign—will likely perform in real life.

A short list of fields rushing to capitalize on these techniques includes advertising, marketing, and promotion; retail and wholesale sales; product production and distribution; human resources; banking, securities, and investments; education; cancer research; epidemiology; physical fitness; criminology; dating; psychology; astronomy; physics; and virtually every sector of government policy and practice. I'm not kidding when I call this a short list—even literary studies are bowing before the altar of Big Data. As Gary King, the director of Harvard's new Institute for Quantitative Social Science, told the *New York Times*, the "march of quantification" is only beginning. "There is no area that is going to be untouched," he said.[3]

I've noted that disassembling wholes into component parts in order to facilitate their manipulation is one of the defining qualities of the nature of technology. To simulate or analyze various forms of experience digitally, in a programmed series of "on" or "off" bits, may be the ultimate extension of that disassembly. So powerful has digital disassembly become that the mathematical concept from which it sprang, information theory, is now considered by many to be science's Holy Grail, the one unifying theory that finally Explains It All. The theory's most avid acolytes hold that *everything* can be reduced to bits. The universe, we're told, is nothing more or less than a cosmic computer, and what we think of as reality may be nothing more or less than a cosmic simulation. I'll return to that later.

It was during the Industrial Revolution that the thrust toward abstraction became conspicuous. Early encounters with the railroad provoked in novice riders an almost psychedelic breakdown of the visual field. Emerson on his first train ride felt he'd realized his lifelong ambition of seeing life in a "panoramic mode": "Matter is phenomenal whilst men & trees & barns whiz by you as fast as the leaves of a dictionary. . . . The very permanence of matter seems compromised & oaks, fields, hills, hitherto esteemed symbols of stability, do absolutely dance by you." Victor Hugo described identical impressions in a letter to his daughter: "The flowers by the side of the road are flowers no longer but flecks, or rather streaks, of red or white; there are no longer any points, everything becomes a streak; the grainfields are great shocks of yellow hair; fields of alfalfa, long green tresses; the towns, the steeples, and the trees perform a crazy mingling dance on the horizon; from time to time, a shadow, a shape, a spectre appears and disappears with lightning speed behind the window; it's a railway guard." It's no surprise that railroad scenes were favorite subjects of the impressionist painters, or that their successors would abandon representational art altogether. "The life of today's cultured person turns more and more away from nature," said the Dutch painter Piet Mondrian; "it is an increasingly abstract life."[4]

Oliver Wendell Holmes claimed that three technologies would free us from the surly bonds of Earth: the stereoscope, the daguerreotype, and the photograph. Their introduction inaugurated, he declared in 1859, "a new epoch in the history of human progress." In this new epoch the physical world could be reproduced so accurately that before long we might not need it anymore. "Form is henceforth divorced from matter," he said. ". . . Matter in large masses must always be fixed and dear; form is cheap and transportable. We have got the fruit of creation now, and need not trouble ourselves with the core."[5]

While many found the perceptual shifts of the industrial era exhilarating, many others found them disorienting. Demarcations once sharply defined were beginning to blur. "With the tremendous acceleration of life, mind and eye have become accustomed to seeing and judging partially or inaccurately," Friedrich Nietzsche said, "and everyone is like the travelers who get to know a land and its people from a railway carriage." Nietzsche also wrote of the "weightlessness" of contemporary culture during this period, while Baudelaire described a "city full of dreams" where ghosts confronted passersby in broad daylight. It was Karl Marx, however, who articulated what may be the defining perception of the modern condition: "All that is solid melts into air."[6]

A parallel movement toward abstraction was taking place at the same moment in mathematics, the field that would give birth to the digital age. According to the historian Israel Kleiner, a "gradual revolution" over the course of the nineteenth century introduced a host of new concepts—set theory, non-Euclidean geometries, noncommutative algebras, space-filling curves, completed infinities—that served to free proofs from any direct connection to the real world. "Mathematicians turned more and more for the genesis of their ideas from the sensory and empirical to the intellectual and abstract," Kleiner said. Suddenly it wasn't what the elements of a given theory consisted of but the relationships between them that mattered.[7]

These developments caused dissension among mathematicians; arguments ensued regarding what constituted a proof. "I believe that the numbers and functions of analysis are not the arbitrary product of our minds," said the French mathematician Charles Hermite, whose work on transcendental numbers helped pave the way for subsequent theories that horrified him. "I believe that they exist outside of us with the same character of necessity as the objects of objective reality; and we find or discover them and study them as do the physicists, chemists and zoologists." Hermite's student, Henri Poincaré, was equally offended. "Logic sometimes makes monsters," Poincaré complained in 1899. "For half a century we have seen a mass of bizarre functions which appear to be forced to resemble as little as possible honest functions which serve some purpose."[8]

There was, of course, no turning back. In 1905 Albert Einstein published his papers on the photoelectric effect and special relativity, heralding another revolution Poincaré found difficult to accept. Einstein had his own problems with abstraction. His biographer Walter Isaacson says that Einstein's greatest gift may have been his ability to extrapolate general theories from the stuff of day-to-day experience. For Einstein the stuff of everyday experience included the clock towers and trains he routinely encountered on his way to and from work as a patent clerk in Bern. The abstraction of math, on the other hand, was something he struggled with all his life, especially as he confronted the implications of quantum mechanics. In his quest for a unified field theory, Einstein found himself relying on ever more abstract mathematical equations, equations he tried in vain to relate to something tangible in nature. "It's like being in an airship," he wrote to a friend, "in which one can cruise around in the clouds but cannot see clearly how one can return to reality, i.e., earth."[9]

The conviction that mathematics is the ultimate reality goes back at least as far as Pythagoras ("Number is all," he supposedly said) and has had any number of prominent proponents ever since, among them René Descartes, who believed mathematics was the key to resolving the ambiguities of philosophy. David F. Noble has written that for Descartes, sense experience was a source of confusion, a problem he solved by turning disembodiment into an explicit epistemological strategy. The mind was of God, the body a product of the Fall, an enemy of reason. Thus he found it necessary, Noble said, to "cleanse" the mind by disconnecting it from fleshly impurity. "Though philosophers had long lamented the liabilities which the body posed for the mind, none before Descartes had actually defined the two as radically distinct and mutually exclusive. In so doing, he aimed to emancipate the divine part of man from its mortal trappings."[10]

It's not surprising that a reverence for numbers would be found at the very origins of computer programming. Ada Lovelace, who in 1843 wrote what many believe was the first computer algorithm (for Charles Babbage's Analytical Engine), called mathematics "the instrument through which the weak mind of man can most effectually read his Creator's work." The lingua franca of the digital age was discovered by George Boole, who in 1854 published his rules of binary algebra, the mathematical logic that forms the foundation of computer programming. Boole shared with Descartes and Lovelace (and with Galileo and Newton) the belief that mathematics was humankind's link with the divine. It was a search for "a primal unity" in nature, David Noble says, that drove Boole's search for that link, and to his formulation of binary algebra. Boole was among the early theorists to endorse the idea that it was relationship rather than meaning that mattered in mathematics. "The validity of the processes of analysis does not depend upon the interpretation of the symbols which are employed, but solely upon the laws of their combination," he wrote.[11]

Just over a century later, a student named Claude Shannon wrote a master's thesis at MIT demonstrating that Boolean logic could be applied mechanically by using the on/off capabilities of telephone routing switches to solve algebraic problems. Shannon subsequently published a two-part paper setting forth the fundamental principles of information theory. Consistent with Boole, he insisted that the only thing that mattered was the relationship between the symbols used to communicate. Any meaning that was subsequently attached to those symbols was, from an engineering standpoint, irrelevant.[12]

Shannon's work, together with Norbert Wiener's theories of cybernetics and the arrival of the microprocessor, set the digital revolution in motion, inspiring a new rash of pronouncements that physicality had given way to abstraction. "The central event of the 20th century is the overthrow of matter," declared a group of high-profile conservatives in a 1994 document modestly titled "Cyberspace and the American Dream: A Magna Carta for the Knowledge Age." Their fustian rhetoric was more than matched by the "Declaration of Independence of Cyberspace," in which a self-declared citizen of that nonlocality, a Grateful Dead lyricist named John Perry Barlow, addressed himself to the "Governments of the Industrial World," which he described as "weary giants of flesh and steel."[13]

"On behalf of the future, I ask you of the past to leave us alone," Barlow wrote. "You are not welcome among us. You have no sovereignty where we gather.... Your legal concepts of property, expression, identity, movement, and context do not apply to us. They are all based on matter, and there is no matter here."[14]

Statements such as these can now be seen as inchoate expressions of the quasi-religious ideology that has attached itself to information theory. The conviction that it all comes down to bits has been steadily gaining momentum since Claude Shannon's and Norbert Wiener's seminal papers appeared in the 1940s and early 1950s. Key to its advance were the breakthroughs in discerning the function of DNA, which suggested that information processing was the essence not only of electronics but also of biology. Add to that the ongoing mysteries of particle physics and the bounteous applications of the digital microprocessor and you have the constituent ingredients of a doctrine that purports to explain, ultimately, existence itself.

Here's a representative statement of the faith from John Brockman, a literary agent who has represented some of information theory's leading exponents, Richard Dawkins and J. Craig Venter among them. "The metaphors of information processing and computation are at the center of today's intellectual action," Brockman said.

> A new and unified language of science is beginning to emerge.... Concepts of information and computation have infiltrated a wide range of sciences, from physics and cosmology, to cognitive psychology, to evolutionary biology, to genetic engineering. Innovations such as the binary code, the bit, and the algorithm have been applied in ways that reach far beyond the programming of computers, and are being used to understand such mysteries as the origins of the universe, the operation of the human body, and the workings of the mind.[15]

On another occasion Brockman added that, thanks to information theory, "we are witnessing a point in which the empirical has intersected with the epistemological: everything becomes new, everything is up for grabs."[16]

Among the things up for grabs is the fundamental character of the human body and of life itself. The idea that the body is a machine has had its share of adherents and detractors over the centuries, but for advocates of information theory, the debate is over. In January 2008, Craig Venter and Richard Dawkins appeared onstage at a Digital Life and Design conference in Munich, Germany. John Brockman moderated the discussion. Here is how Venter described his work and philosophy:

> For the past 15 years, we have been digitizing biology. When we decoded the genome, including sequencing the human genome, that's going from what we consider the analog world of biology into the digital world of the computer. Now, for the first time, we can go in the other direction. With synthetic genomics and synthetic biology, we are starting with that purely digital world. We take the sequence out of the computer and we chemically, from four raw chemicals that come in bottles, can reconstruct a chromosome in the laboratory, based on either design, copying what was in the digital world, or coming up with new digital versions.[17]

At that point Brockman broke in to ask Venter if by using the word "design" he meant to imply that life is a technology. "Life is machinery," Venter replied. "Life becomes a form of technology as we learn how to engineer and reproduce it."[18]

Richard Dawkins was, if anything, more emphatic. The genetic code, he said,

> is pure information. You could put it into a printed book. You could send it over the Internet. You could store it on a magnetic disk for a thousand years, and then in a thousand years' time, with the technology that they'll have then, it will be possible to reconstruct whatever living organism was here now. So, this is something which was utterly undreamed of before the molecular information revolution. What has happened is that genetics has become a branch of information technology. It is pure information. It's digital information. It's precisely the kind of information that can be translated digit for digit, byte for byte, into any other kind of information and then translated back again. This is a major revolution. I suppose it's probably *the* major revolution in the whole history of our understanding of ourselves.[19]

For the transhumanists, information is to the key to transformation. Ray Kurzweil, for example, bases many of his predictions on our ability to build increasingly sophisticated models of the human brain. These simulations will become so accurate, he believes, that we will be able to duplicate human consciousness in machines, where it can be run at super speeds like any other software program. "A key observation regarding the Singularity," he says, "is that information processes—computation—will ultimately drive everything that is important."[20]

Readers who have come with me this far won't be surprised that I greet such claims with skepticism, again, not because I doubt they'll come true, but because I don't think those who make them adequately take into account their implications. Evidence suggests that the rush to exploit data and simulation technologies is following the same pattern that pretty much every other technological breakthrough of the past two centuries has followed: Discovery results in exhilaration bordering on delirium. Morning-after qualifications are dutifully issued and promptly ignored. The rush to exploitation proceeds, unexpected consequences trailing in its wake.

The problem with Big Data and the models based on Big Data is that they are facsimiles of reality—slices of reality, not reality itself. That's precisely why they're useful. Models allow us to see the elements we want to focus on more clearly because they help us eliminate material that, for purposes of the present analysis, we deem extraneous. The philosopher and historian of science Georges Canguilhem put it well: "A model only becomes fertile by its own impoverishment."[21]

Even when useful, though, impoverishment remains impoverishment. Real life is more complicated and richer, rife with ambiguity, uncertainty, and qualification. This was the point of the general semantics scholar Alfred Korzybski, who famously said, "The map is not the territory." The truth of this has not escaped the more perceptive advocates of Big Data and simulations, which is why they regularly insist that what those techniques tell us must continually be verified and augmented by human judgment. This is fine as far as it goes. Frequently it doesn't go far enough, however, precisely because the thrust of those technologies—their reason for being—is the *replacement* of human judgment. It's a built-in contradiction, and the power of the technologies employed constantly threatens to seduce or bully the human being into submission. As Rachel Schutt, a senior statistician at Google Research (she subsequently moved to the role of Chief Data Scientist at News Corporation), told the *New York Times*, "Models do not just predict, but they can make things happen. That's not discussed generally in our field." A student studying simulation technologies echoed that sentiment. "It is easy to let the simulation manipulate you instead of the other way around," she said.[22]

As useful as models can be, then, they can also be misleading, and therefore dangerous. Anyone who doubts that only needs to study the roots of the global financial crisis of 2008, a crisis that hobbled economies and caused serious, often irreparable damage to tens of millions of businesses, families, and individuals. That crisis stemmed directly from the investment community's use of financial techniques—bundled mortgage securities and sophisticated trading algorithms—that removed traders' personal judgment from what became, for a time, incredibly powerful engines of (paper) profit. Numerous books have documented how Wall Street bankers were lured by greed into believing what their algorithms told them and what their mortgage bundles obscured, ignoring warnings from those who recognized that the emperor had no clothes. The titles of two of those books capture the essence of what went wrong: *Models Behaving Badly* and *A Colossal Failure of Common Sense*.[23]

As I mentioned, the potential applications of data mining and simulation are so broad as to be almost all-encompassing, and in every arena where something of value can be gained, something of value can also be lost. Sherry Turkle cites the story of sociologist Paul Starr, who worked for the Bill Clinton administration on health care reform. Starr noticed what he considered a mistaken assumption in a simulation model that the Congressional Budget Office was likely to use in the debate and criticized it. A colleague told him not to waste his breath. The CBO would never back off the dictates of its model, the colleague said; policy would have to adjust. Turkle calls this a case of "simulation resignation."[24]

The dynamics of simulation resignation are such that what begins as an abstraction becomes an embodiment that replaces the more abundant embodiment it was intended to represent. The novelist Ellen Ullman offered a powerful description of this process in *Close to the Machine*, her memoir of her days as a computer programmer.[25]

In the book's opening chapter, Ullman describes a meeting she has with a group of clients for whom she and her colleagues are designing a computer system, one that will allow AIDS patients in San Francisco to deal more smoothly with the various agencies that provide them services. Typically, this meeting has been put off by the project's initiating agency, so that the system's software is half completed by the time Ullman and her team actually sit down with the people for whom it is ostensibly designed.

As the meeting begins, it quickly becomes apparent that all the clients are unhappy for one reason or another: the needs of their agencies haven't been adequately incorporated into the system. Suddenly, the comfortable abstractions on which Ullman and her colleagues based their system begin to take on "fleshly existence." That prospect terrifies Ullman. "I wished, earnestly, I could just replace the abstractions with the actual people," she writes. "But it was already too late for that. The system pre-existed the people. Screens were prototyped. Data elements were defined. The machine events already had more reality, had been with me longer, than the human beings at the conference table. Immediately, I saw it was a problem not of replacing one reality with another but of two realities. I was there at the edge: the interface of the system, in all its existence, to the people, in all their existence."[26]

The real people at the meeting continue to describe their needs and to insist they haven't been accommodated. Ullman takes copious notes, pretending that she's outlining needed revisions. In truth she's trying to figure out how to save the system. The programmers retreat to discuss which demands can be integrated into the existing matrix and which will have to be ignored. The talk is of "globals," "parameters," and "relational database normalization." The fleshly existence of the end users is forgotten once more. Ullman says,

> Some part of me mourns, but I know there is no other way: human needs must cross the line into code. They must pass through this semipermeable membrane where urgency, fear, and hope are filtered out, and only reason travels across. There is no other way. Real, death-inducing viruses do not travel here. Actual human confusions cannot live here. Everything we want accomplished, everything the system is to provide, must be denatured in its crossing to the machine, or else the system will die.[27]

Ultimately Ullman finds herself reassessing one of our most treasured assumptions regarding technology. "I'd like to think that computers are neutral, a tool like any other," she says, "a hammer that can build a house or smash a skull. But there is something in the system itself, in the formal logic of programs and data, that recreates the world in its own image."[28]

It's tempting to avoid fleshly existences if we can; simulations aren't as messy. This is true even of those who deal with fleshly existences for a living, such as doctors. The lab tests and imaging technologies available to physicians today have saved the lives of countless patients, and the accumulation of those results is providing invaluable insights into the etiology of disease. At the same time, however, their efficiency is encouraging many doctors to lose touch, literally, with their patients. Abraham Verghese, a professor at the Stanford University School of Medicine, confessed in an op-ed piece in the *New York Times* that in his own practice he often found himself more focused on the data in a patient's computerized medical records than on the patient himself. The computer record creates what Verghese calls an "iPatient," and it is the iPatient who becomes "the real focus of our attention, while the real patient in the bed often feels neglected, a mere placeholder for the virtual record."[29]

Verghese tells the story of one patient who arrived in the emergency room with severe seizures and breathing difficulties. A CT scan revealed that she had cancerous tumors in both breasts and throughout her body. It was clear from the scanned images, Verghese says, that the tumors in the patient's breasts could have been detected far earlier by a simple physical exam. Although she'd been seen several times previously by several different physicians for less severe symptoms—symptoms probably related to her cancer—they never were.

I could go on citing examples—the prostitution of high school education to testing models is one that's especially infuriating—but I think the point is made. I can't resist, though, adding one quote from the Air Force general who was in charge of investigating how U.S. helicopters, acting on information supplied by a drone video operator seven thousand miles away in Nevada, attacked a convoy of unarmed civilians in Afghanistan in February 2010. Depending on whose figures you believe, the attack killed between fifteen and twenty-three men, women, and children. "Technology can occasionally give you a false sense of security that you can see everything, that you can hear everything, that you know everything," said Air Force Maj. Gen. James O. Poss.[30]

I suspect, though I'm not certain, that the intoxication that characterizes the practical applications of data and simulation may also be evident in the more theoretical realms of information theory. Specifically I wonder what it actually means to say that information is the fundamental property of the universe. When the science writer James Gleick calls information "what our world runs on: the blood and the fuel, the vital principle," is he describing something tangible, or is it, as John Brockman said, a metaphor that can be applied to an almost infinite variety of processes and relationships? For me the answer is never entirely clear.[31]

A long string of terms has been applied through history to describe the irreducible property of existence, many of them blurring the line between metaphor and reality. Water, air, fire, number, energy, atoms, and organism have had their moments, some repeatedly. More mechanical images—electricity, steam, the motor, force—came into fashion with the Industrial Revolution. Thus it's not surprising that in an age of information, information would suddenly be perceived as the single, all-purpose element underlying everything.[32]

It's also not entirely clear exactly where our trajectory toward abstraction might take us, morally as well as practically. If it all comes down to bits, is there any difference between nature and technology? Craig Venter and many others think not. Descartes agreed, to a point—in his more religious times, he wasn't ready to go quite that far. Animals are machines, he said, but not human beings. A century later Julien Offray de la Mettrie dared to take the next logical step. "Let us then conclude boldly that man is a machine," he wrote, "and that in the whole universe there is but a single substance differently modified."[33]

In her brilliant book, *How We Became Posthuman*, N. Katherine Hayles describes Claude Shannon's view of information as "a probability function with no dimensions, no materiality, and no necessary connection with meaning . . . a pattern, not a presence." Such a definition, she says, cuts information loose not only from meaning but also from embodiment. "Abstracting information from a material base meant that information could become free-floating, unaffected by changes in context. The technical leverage this move gained was considerable, for by formalizing information into a mathematical function, Shannon was able to develop theorems, powerful in their generality, that hold true regardless of the medium in which the information is instantiated."[34]

This helps explain why defining what information is, exactly, seems so elusive. It also explains its appeal. Information can be all things to all people while it achieves, at long last, the overthrow of matter. "From here," Hayles writes, "it is a small step to perceiving information as more mobile, more important, more *essential* than material forms. When this impression becomes part of your cultural mindset, you have entered the condition of virtuality."[35]

Late in his career, Marshall McLuhan foresaw the emergence of the condition of virtuality, and it caused him to reassess the optimism he'd earlier expressed regarding "the global village." Instead he began to speak of "discarnate man," a creature his biographer, Philip Marchand, describes as

> the electronic man, the human being used to talking to other humans hundreds of miles away on the telephone, used to having people invade his living room and his nervous system via the television set. Discarnate man had absorbed the fact that he could be present, minus his body, in many different places simultaneously, through electronics. . . . His self was no longer his physical body so much as it was an image of a pattern of information, inhabiting a world of other images and other patterns of information.[36]

The effect of this reality was to give discarnate man an overwhelming affinity for, in McLuhan's words, "a world between fantasy and dream" and a "typically hypnotic state," in which he was totally involved in the play of images and information, like a small child fascinated by a kaleidoscope.[37]

There's irony in the current excitement over information because both Claude Shannon and Norbert Wiener worried about the spread of their ideas into domains where they didn't belong. Wiener, schooled in philosophy as well as science, indulged his own speculations on the broader implications of cybernetics, many of them gloomy, as I'll discuss later. Shannon remained circumspect. At one point he went to the trouble of publishing an editorial warning that a "bandwagon" effect had overtaken the nascent field. As a result, he said, information theory had "ballooned to an importance beyond its actual accomplishments. . . . Our fellow scientists in many different fields, attracted by the fanfare and by the new avenues opened to scientific analysis, are using these ideas in their own problems. Applications are being made to biology, psychology, linguistics, fundamental physics, economics, the theory of organization, and many others. In short, information theory is currently partaking of a somewhat heady draught of general popularity."[38]

Shannon conceded that information theory might, in time, prove useful in many of these fields, but that would have to happen not through speculation but through "the slow tedious process of hypothesis and experimental verification."[39]

N. Katherine Hayles points out that not even the founders of a theory can control the interpretations it accumulates once it begins to spread through the culture "by all manner of promiscuous couplings." In the case of information theory, so promiscuous have those couplings become that I could have left off the "quasi" in my earlier reference to its "quasi-religious" status. The virtual reality pioneer Jaron Lanier has written that the belief system he calls cybernetic totalism is a pervasive, almost inescapable orthodoxy in Silicon Valley. He also agrees with Hayles that such a view all too easily becomes untethered from embodiment. "Once you have subsumed something into its cybernetic reduction," Lanier says, "any particular reshuffling of its bits seems unimportant."[40]

All that is solid melts into air.

PART IV

BOUNDARY ISSUES

*May not man himself become
a sort of parasite upon the
machines? An affectionate
machine-tickling aphid?*

SAMUEL BUTLER

CHAPTER 13

SHAPERS SHAPED

To start with, we are what our world invites
us to be, and the basic features of our soul are
impressed upon it by the form of its surroundings
as in a mold. Naturally, for our life is no other
than our relations with the world around.

JOSÉ ORTEGA Y GASSET

The anticipation transhumanists feel for the coming merger of human and machine sometimes takes on a slightly perverse quality. In *The Singularity Is Near*, for example, our old friend Ray Kurzweil imagines the spectacular varieties of sexual experience that await us once we inject virtual reality nanobots into our brains.

With minds so adapted, Kurzweil says, we'll be able to have sex with whomever we want to have sex with, from Marilyn Monroe to Brad Pitt (or both together, if you so desire), wherever we want (the royal bedroom at Versailles, perhaps, or the fifty-yard line at the Super Bowl), however we want (with virtual bodies, who knows what's possible?). We'll also be able to project whatever physique and personality we'd like to present to our lovers, aware that they're able to do the same, meanwhile imagining that they're having sex with . . . whomever they want to have sex with. Of course, when unlimited sexual experience can be ours at any time of the day or night, just by thinking about it, a "real life" partner seems somewhat superfluous.

This captures the essence of the technological dream: endless stimulation and satisfaction, on demand, forever. The thinking seems to be, if it's technically possible, why settle for less?

So eagerly are these developments anticipated in fact, that a transhumanist movement has formed, replete with manifestos and organizations to further the cause. Why it's necessary to promote a transformation that is seen by most of its advocates as inevitable, I'm not sure. Probably the thinking is that, since our transformation through technology will create a race of superior beings, and since it's obvious the human species is, at its present state of development, far from perfect, the sooner we get there the better.

As I've mentioned, a fundamental tenet of transhumanism is that the merging of humans and machines is the next step in the long upward march of evolution, and that therefore human beings as we now know them will become—are in the process of becoming—the latest casualties of natural selection. A corollary of that argument is a deep-seated contempt for the biological bodies natural selection has given us so far, which is one reason virtual sex seems so much more alluring than old-fashioned sex.

"Transhumanists view human nature as a work-in-progress, a half-baked beginning that we can learn to remold in desirable ways," says Nick Bostrom of the Institute for the Future of Humanity at Oxford University. "Current humanity need not be the endpoint of evolution. Transhumanists hope that by responsible use of science, technology, and other rational means, we shall eventually manage to become posthuman beings with vastly greater capacities than present human beings have."[1]

It's striking how often these sorts of dismissive comments about the human body, simultaneously prideful and self-loathing, come up in the transhumanist literature. Kurzweil writes that our "version 1.0. biological bodies" are "frail and subject to a myriad of failure modes, not to mention the cumbersome maintenance rituals they require." He concedes that the designs of nature reflect "ingenuity" but quickly adds that they remain "many orders of magnitude less capable than what we will be able to engineer." Kurzweil's mentor at MIT, Marvin Minsky, has called the body a "bloody mess of organic matter." No wonder transhumanists await the Singularity with such impatience. Their own bodies fill them with disgust.[2]

The point here isn't to argue whether the transhumanist agenda will or won't materialize. Again, betting against technological advance is a fool's game. I do question the underlying assumption that greater technological prowess automatically leads to greater happiness, but such sentiments will strike the transhumanists as just that: sentiments, and thus as useful as an appendix.

The truth is I feel oddly conflicted about the legitimacy of the transhumanist claim that humans will inevitably merge with their machines. In some ways I agree that technology's absorption of the human mind and body is well under way. Indeed, there are plenty of reasons to think it's further along in more ways than the transhumanists typically have in mind. At the same time I see the integration of human being and machine to be an experiment doomed for failure and a travesty in the making, pretty much along the lines that Mary Shelley laid out almost two centuries ago.

I've said that technology has a natural tendency to expand its sphere of influence, displacing the natural world as it does. Given that basic principle, there's no reason to suspect that human beings would be more immune from technological imperialism than any other natural entity. To the contrary, considering all we've learned about how sensitive humans are to their environment, it would be surprising if we *weren't* profoundly affected by a force as powerful and all-encompassing as technique.

I need here to restate my conviction that we err if we think of technology only in terms of computers, the Internet, cell phones, and other digital devices, as we are wont to do. We must include in our thinking *all* the artifacts and systems that constitute the technological society, from the mundane (candy bars and clock radios) to the monumental (skyscrapers and nuclear bombs). It is through the influence of technology *in its entirety* that the merging of man and machine may have been, to a degree we may not appreciate, already accomplished.

I base that statement in part on the evidence provided by some of the more dramatic breakthroughs in the biological sciences in recent decades. The collective weight of a host of studies has confirmed that the human body is a dynamic rather than a static entity, one that responds more sensitively and in more significant ways to environmental influences than we ever suspected. (The "we" I refer to in that sentence is Western culture in general and Western science in particular. Various "primitive" cultures have for centuries assumed an intimate relationship between humans and their environment, although in their case the environment has been natural rather than human-made.) This is a subject that's been written about a lot lately, so I'll just touch on three key discoveries.

The first is neuroplasticity, which refers to the fact that the human brain—in adults as well as children—literally rewires itself in response to what we experience. The fact that the brain continues to re-form itself throughout life overturns the longstanding assumption that its development ends at the age of five or so. We now know that existing neuronal circuits can be strengthened or weakened and new circuits formed by experiences both voluntary (studying, for example) and involuntary (such as stress).[3]

Recent breakthroughs in genetics have similar implications. Since Darwin it's been assumed that evolutionary changes take many generations to unfold. In the past twenty years researchers have discovered that while altering DNA is a long-term proposition, altering how DNA is actually expressed within an organism is not. Genomes have markers that determine whether our genetic predispositions are manifested. Those markers can be switched on or off by environmental influences, making it more likely, for example, that we'll develop diabetes or less likely that we'll be fast learners. By flipping that genetic switch, environmentally provoked changes in gene expression can be inherited within a single generation and can be passed on to at least four subsequent generations. As a report in *Time* magazine put it, "The more we lift the lid on the genome, the more vulnerable to experience genes appear to be."[4]

We've also become increasingly aware in recent decades that we're an organism with two brains, not one: an electrical brain that operates through the nervous system and a chemical brain that circulates through every cell in the body. Molecules called peptides are the messengers of that chemical brain, which is considerably older and more basic than its electrical counterpart. Produced and secreted by organs ranging from the pituitary gland to the kidneys, peptides help regulate everything from appetite and orgasm to breathing and bowel movements. These chemicals, says Candice Pert, a neuroscientist who was instrumental in discovering their role, "weave the body's organs and systems into a single web that reacts to both internal and external environmental changes with complex, subtly orchestrated responses."[5]

After I'd written that last paragraph, I showed it to a friend of mine, Roger Cubicciotti, to see if he thought I'd gotten it right. Roger has a PhD in biology and a string of patents in nanotechnology, many of them involving ways in which nanobots can be injected into the body to detect, diagnose, and treat illness. He said what I'd written was okay as far as it went, but it didn't go far enough. My comment that the body has two brains, one chemical and one electrical, vastly underplays the complexity, and the wonder, of the situation, he said. In truth our physical selves constitute a "multidimensional coordinate system." Each of us has within us literally dozens of classes of molecules that interact in hundreds of different ways with hundreds of thousands of receptors, effectors, and neural networks. Those internal dynamics, in turn, respond in countless ways to the multitude of influences coming into the body from the outside—some natural, others man-made. The number of possible interactions produced by this combination of influences and potential responses is virtually infinite. This is especially true because when disparate systems interact, "emergent"—meaning unpredictable—properties can arise. What all this adds up to is a degree of sensitivity that is intricate beyond comprehension.

There's one other point I'd like to touch on regarding the intimacy we share with machines. It involves relationality at the subatomic level. Quantum physics tells us that there is an effective relationship between every particle of the cosmos. "In the modern concept," said Alfred North Whitehead, "the group of agitations which we term matter is fused into its environment. There is no possibility of a detached, self-contained local existence. The environment enters into the nature of each thing." What influences this subatomic relationality might have on our daily existence is as yet unknown, but it would be rash to assume influences don't exist.[6]

As I say, none of this is especially new, but I needed to review some of the basics in order to set up my argument. It's probably obvious by now where I'm heading.

We have, on the one hand, a human organism that is shaped, literally and figuratively, by its environment. On the other hand, we have a technological environment that bears little resemblance to the natural environment in which that organism has evolved and thrived for thousands of years. Technology is in the air we breathe, in the food we eat, and in the water we drink. We absorb technology through our skin. Tests regularly establish that literally hundreds of chemical contaminants are present at any given time in the bodies of every man, woman, and child on the planet—scientists say that babies today are born "pre-polluted." Considering all this, it hardly seems unreasonable to conclude that if we aren't posthuman yet, we're certainly not as human as we used to be.[7]

It's a central conviction of the transhumanists that the connection between human and machine is a natural one, that the merger of human and machine is where evolution has been taking us all along. Andy Clark, author of *Natural Born Cyborgs*, believes humans have a genetic predisposition for uniting with material objects, a predisposition that will ultimately free us from the inherent limitations of what he likes to call our "biological skinbags." "We have been designed, by Mother Nature, to exploit deep neural plasticity in order to become one with our best and most reliable tools," Clark says. "Minds like ours were made for mergers. Tools-R-Us, and always have been."[8]

It's true that over time we've consistently projected ourselves into our technologies, almost literally. Many of our earliest tools were modeled on our bodies: the shape of a bowl mirrored the shape of two cupped hands; a spear lengthened the reach of an arm. "Man is a shrewd inventor and is ever taking the hint of a new machine from his own structure," observed Emerson, "adapting some secret of his own anatomy in iron, wood and leather to some required function in the work of the world."[9]

The word "prosthesis" derives from the Greek *prosthenos*, for "extension." The first modern writer to explore this connection systematically was probably Ernst Kapp in 1877. Kapp speculated that railroad systems unconsciously mimicked the circulatory system, while telegraph lines extended the nervous system. Marshall McLuhan expounded the same idea in *Understanding Media*, which bears the subtitle *The Extensions of Man*. "It is a persistent theme of this book," McLuhan wrote, "that all technologies are extensions of our physical and nervous systems to increase power and speed."[10]

The idea that technology represents a natural extension of the natural order has long been employed as an argument in favor of technological development. It was just such an argument, in fact, that opened the way for the industrialization of America in the earliest days of the republic. As tempted as they were by the riches of industry, citizens of the newly formed union did not surrender the agrarian ideal without misgivings, especially in light of the human exploitation and environmental degradation that characterized the industrial experience in England. One of those who helped overcome those fears was Tench Coxe, an assistant to Secretary of the Treasury Alexander Hamilton.

Coxe wrote the first draft of Hamilton's *Report on Manufactures*, which enthusiastically endorsed industrialization as the road to prosperity and freedom. A tireless promoter of the new technologies, Coxe argued that the abundance of our natural resources would save us from repeating England's mistakes. Factories would fit *into* the land, he said, but wouldn't overwhelm it—a vision of harmony between nature and technology that Leo Marx would describe as "the machine in the garden." Coxe underscored that vision by insisting that, like nature, technology was a manifestation of God's creation. The United States was blessed, he said in a speech in 1797, with "fruitful soil," "healthful climate," "mighty rivers and adjacent seas abounding with fish. . . . Agriculture, manufactures and commerce, *naturally arising* from these sources, afford to our industrious citizens certain subsistence and innumerable opportunities of acquiring wealth."[11]

The same theme would be frequently sounded in support of Manifest Destiny. Technological development was perceived as the "legitimate offspring" of forests, mountains, and rivers, said the historian Perry Miller, "not a violation of Nature but an embrace." Miller quoted a nineteenth-century testament to the transformative power of the railroad. "Concentration of thought, purpose, will, means, and men," it read, "—not futile and impotent, but quick with life and taking shape in action, and that action tending, not to rebuild and perpetuate the old and decayed, nor to hem in what is, so that it should never be aught else, but to transform it into something better, and in the transformation to make it give forth new qualities, and put forth new and more exquisite beauties."[12]

We can see in statements such as these the beginnings of the mechanistic animism that would evolve into transhumanism.

I for one don't question that we humans have a tendency to develop a sense of connection with our tools. As an amateur guitarist, I feel a strong attachment to the Martin D-18 I've played for thirty years. I've also felt myself merging with my car on long road trips, when after several hours it feels more natural sitting behind the wheel than it does walking around. Witness, too, our attitude as drivers toward pedestrians, whom we tend to consider annoying obstacles impeding our momentum. The human-machine complex brooks interference with its forward motion only grudgingly.

The question is whether we can *literally* merge with our machines, and if we can, how healthy such a merger would be. The transhumanists are sure we're well on our way to becoming fully integrated cyborgs and that we'll be much better off once that fusion is complete. Others say that's absurd, that humans are biological organisms of flesh and blood and so of a fundamentally different order of being than machines, and to suggest otherwise is to propose an abomination.

Jacques Ellul was in the latter group. He didn't disagree that a merger of human and machine was under way, but he considered that process not a felicitous melding but a takeover by a superior force. The outcome would be "a profound mutation," he believed, one that required the "dismembering and complete reconstitution" of the human being. "The machine tends not only to create a new human environment, but also to modify man's very essence," Ellul wrote. "... When technique enters into every area of life, including the human, it ceases to be external and becomes his very substance. It is no longer face to face with man but is integrated with him, and it progressively absorbs him."[13]

Because the breakthroughs in biology I've mentioned are very new, the idea that we might be transformed in some intrinsic way by our interaction with machines has mostly been contemplated in metaphorical, spiritual, or psychological terms. Nonetheless, it's remarkable how perceptively some artists have sensed the degree to which that transformation might be taking place. Mary Shelley again comes to mind, as does Thomas Carlyle. "Let us observe," Carlyle wrote in 1829, "how the mechanical genius of our time has diffused itself into quite other provinces. Not the external and physical alone is now managed by machinery, but the internal and spiritual also. ... Men are grown mechanical in head and in heart, as well as in hand."[14]

Among those Carlyle influenced, and who penned similar observations of their own, were Emerson ("Tools are in the saddle, / and ride mankind") and Thoreau ("men have become tools of their tools"). Another writer who expressed remarkably similar opinions in a radically different style was Karl Marx. The factory, Marx said, turns workers into "a mere appendage" of the machine so that they become "parts of a living mechanism."[15]

That statement calls to mind the iconic image of Charlie Chaplin's Little Tramp being threaded through the giant wheels of the factory machine in *Modern Times.* Unless you've watched the film recently, you may have forgotten an earlier scene in which Chaplin's whole body violently jerks and twitches in rhythm with the assembly line even after he's stopped working on it. He'd absorbed, or been absorbed by, the heartbeat of the machine. He'd become a cyborg.[16]

Moviegoers who saw *Modern Times* when it was first released recognized that one of its targets was Henry Ford, progenitor of the assembly line Chaplin depicted. Ford, whose standing with the public had by then declined, agreed with Chaplin that modern times had grown radically out of balance, although from his perspective the city was the source of the distortion. Like so many before him, he believed the answer was a harmonious combination of technology and nature. Ford envisioned men spending part of their time working farms and the rest in small factories—"village industries"—that would free them from urban imprisonment. "I think normal life for a man is to get back to the land," he said in 1926. "The land is the healthiest place to be. The trouble up to now is that a man couldn't get enough experience on the farm; but now, with the telephone, the phonograph, the moving pictures, the automobile . . . the farmer can live in the country and have all the experience in the world."[17]

We know now that Ford's confidence on that score was premature. The few village industry sites he did establish failed. Subsequently the suburbs promised to become the perfect bridge between city and nature, but that doesn't seem to have worked out either. We were told that with the Internet we'd finally be free to live in the country and have all the experience in the world, but finding a way to get paid for it has proved harder than expected, and farming has been industrialized, too. Factories, meanwhile, are bigger than ever, although fewer of them are in America. We're still waiting for the overthrow of matter.

CHAPTER 14

ECOTONE

> He hangs between; in doubt to act or rest
> In doubt to deem himself a god or beast
>
> ALEXANDER POPE

I've mentioned several times the belief of Ray Kurzweil and his fellow transhumanists that humankind is well on its way to the Singularity, when our ongoing merger with machines will be completed. I find it interesting that at the same time the popular imagination is increasingly filled with thoughts of cyborgism, there's been an unusual amount of discussion about where the boundary lies between humans and animals. The fact that these two conversations are taking place in parallel is not, I think, a coincidence.

The Minotaur and centaurs of ancient myth and the animal costumes in the rituals of indigenous tribes testify to the fact that, throughout human history, we've imagined the boundary between ourselves and animals as permeable, figuratively if not literally. Darwin raised the stakes on the identity question significantly by making it a scientific issue rather than a mythological construct. Do humans represent a radical break in the evolutionary chain, or do we share more in common with our animal friends than many of us would like to believe?[1]

The answer to that question tends to vary according to who's giving it and when. "There has always been a good degree of tension between views that stress discontinuity and those that stress continuity between humans and animals," writes Raymond Corbey, author of *The Metaphysics of Apes: Negotiating the Animal-Human Boundary*. For much of recent history, Corbey says, the underlying agenda in the debate was to maintain discontinuity in order to preserve both human dignity and the idea of evolutionary progress. More recently we've witnessed a shift toward an emphasis on the commonalities humans share with the animal world.[2]

Much of this shift has been led by science. Researchers keep finding that we share more, in terms of both behavior and physiology, with "lower" species than we ever suspected. One sign of this collective reevaluation is the Cambridge Declaration on Consciousness, issued in 2012 by a prominent group of neuroscientists. The weight of scientific evidence affirms "that humans are not unique in possessing the neurological substrates that generate consciousness," the declaration says. "Non-human animals, including all mammals and birds, and many other creatures, including octopuses, also possess these neurological substrates." These substrates, the declaration adds, have been shown to produce "feeling states" and to provoke "emotional behaviors" in animals as well as humans.[3]

Technology, meanwhile, is raising sensitive new issues regarding the compatibility of animal and human biologies. Lee Silver, a professor of molecular biology at Princeton University, has written of the ethical dilemmas being raised by research in which human stem cells are injected into mice and other laboratory animals. The resulting creatures are mostly animal but partly human. Scientists call them "chimeras," after the monster in Greek mythology that was part lion, part goat, and part snake. The hope is that one day we'll be able to put an end to transplant shortages by growing human organs within pigs and other animals. Silver thinks that's a technique most people will find acceptable; breeding monkeys with human brains might be another story. In any event, the fact that such hybrids are within reach forces us, Silver says, to contemplate whether the existence of a strict line separating human beings from animals "may simply be a figment of our imagination."[4]

This also seems to be a time when people are especially interested in connecting with animals. Dogs and cats are lavished with attention while tour guides and hotels profit by putting people in touch, literally, with whales and dolphins. I'm reminded of Anne Foerst's comment that our need for communion with animals reflects the fact that humans are "a deeply lonely species." This explains, she believes, her readiness to respond emotionally to robots.

"The Family Dog," a video report produced by the New York Times, demonstrates a striking example of that readiness. It focuses on a Japanese husband and wife who were deeply attached to Aibo, their robotic pet dog. Sony sold 150,000 Aibos before discontinuing them in 2006, and many of their owners became distraught at the announcement, knowing it would soon be impossible to repair them. "When I first got Aibo, it was like having a new baby," the woman says. "I can't live without it now." The video opens in a Shinto temple where a funeral service is in progress for a number of Aibos that have passed beyond the point of no return. "The meaning of this Aibo funeral comes from our realization that everything is connected," the priest tells the gathered mourners. "The animate and the inanimate are not separated in this world. We have to look deeper to see this connection. We pray for the spirit which resides inside Aibo to hear our prayers and feelings."⁵

This may sound odd to Western ears, even though we demonstrate, as I've mentioned, our own ability to bond with the machines in our midst. Science fiction suggests we'll feel increasingly comfortable sharing our lives with increasingly sentient technologies. Movies and TV programs often show humans and robots engaged in perfectly natural interactions—the android Data on *Star Trek* comes to mind.

Roboticists share those visions and are actively working to bring them to fruition, intimately. Another *New York Times* video shows a workshop where makeup artists apply the finishing touches to technologically advanced, anatomically correct sex dolls. These playmates are far more lifelike than the inflatable variety that have been the source of raunchy jokes for years. They not only look more realistic; they talk and seem to respond intelligently to questions. Matt McMullen, the developer of the dolls in the video, says his ambition is to create robots that will connect with their human partners personally as well as carnally—most of us, after all, don't consider purely mechanical sex especially satisfying. Much better, McMullen says, if people will be able to "develop some kind of love for this being."⁶

It's an open question how deep such an attachment can go. Sony discontinued the Aibo, we have to assume, because its potential as a mass-market product seemed limited. And the robots we see on TV and in movies are typically portrayed as more threatening than endearing—HAL in Stanley Kubrick's *2001* being the classic example.

The passionate connection we feel for our biological pets is to some degree a response, I believe, to the alienation we feel within our increasingly technological environment. Living in a largely artificial "surround" (as academics describe the physical setting that constitutes our immediate experience) exacerbates our loneliness as a species. Most of us have heard by now of the "uncanny valley," the term engineers use to describe the uneasiness caused by robots that look mostly human but not quite. I suspect that uneasiness derives from a largely unconscious perception that a machine is trying to sneak across the organic-mechanical boundary. On some intuitive level we don't trust it. Contact with animals provides a reassuring relationship with something other than machines and the products of machines.⁷

Transhumanists see this distrust as highly illogical. Again, from their perspective, humans have been merging with their technologies for millennia; that's what separates us from the animals. "We are brothers and sisters of our machines," writes George Dyson, author of *Darwin among the Machines: The*

Evolution of Global Intelligence. "Minds and tools have been sharpened against each other ever since a scavenger's stone fractured cleanly and the first cutting edge was held in a hunter's hand."[8]

At the same time many transhumanists concede that their own comfort with cyborgism may not be shared by the average man and woman on the street. The species as a whole, they say, is dealing with a "fourth discontinuity" in human particularity. The fourth discontinuity theory is based on a comment by Sigmund Freud. Through most of their history, he said, human beings were confident they were at the center of the universe. In the past four hundred years or so, however, that cherished self-image has suffered three major blows. The first of these was the Copernican Revolution, when we learned that Earth is a satellite of the sun, rather than the other way around. The second was Darwin's *On the Origin of Species*, which revealed that man is descended from apes. The third, Freud modestly contended, was his theory of psychoanalysis, which demonstrated that our thoughts and behaviors are not entirely in our control but instead are influenced by drives and conflicts hidden deep within our subconscious.

The fourth discontinuity is a recent addition to the list that takes into account how quickly machine intelligence has advanced in the past fifty years. Just as the second discontinuity acknowledged that we can no longer claim to be an order of nature distinct from and superior to animals, the fourth discontinuity holds that we can no longer claim to be an order of nature that is distinct from and superior to machines. The result is an ongoing identity crisis.[9]

Paul Tillich described a similar sense of insecurity. Like other existentialists, he believed that uneasiness is endemic to the human condition. It's weird being aware that we exist and weird knowing we're going to die. Our predicament leaves us with persistent feelings of, as Tillich put it, "uncanniness." We've come up with lots of ways to avoid those feelings, and technology is high on the list. On one level we find technology reassuring, Tillich said, because we think we can control it. Even though we may not understand how it works, we believe it behaves by rational, logical, "calculable" rules. We can cloak ourselves in it and feel secure. Tillich cited the home as an example. Its "coziness," he said, holds "the uncanniness of infinite space" at bay. What the house or apartment offers individuals, the city offers humans en masse.[10]

Like so many palliatives, however, technology can turn on us. It may not be as safely under control as we'd hoped. The potential for unease grows as our technologies become more powerful, more complex, and more self-determined. On some level we're aware that the relentless logic they're following is their own. We know they're not truly alive, but they seem to be. We wonder whose agenda is being followed. "As the technical structures develop an independent existence," Tillich said, "a new element of uncanniness emerges in the midst of what is most well known. And this uncanny shadow of technology will grow to the same extent that the whole earth becomes the 'technical city' and the 'technical house.' Who can still control it?"[11]

The same shadow had been noticed earlier by Walter Benjamin. "Warmth is ebbing from things," he said. "The objects of daily use gently but insistently repel us. Day by day, in overcoming the sum of secret resistances—not only the overt ones—that they put in our way, we have an immense labor to perform. We must compensate with our warmth if they are not to freeze us to death."[12]

Humans have been trying to figure out where they stand in relation to animals and machines at least since the scientific revolution began. Francis Bacon's secretary, Thomas Hobbes, was an example. His seminal book of political philosophy, *Leviathan*, was published a year after the death of Descartes, who held that animals were machines but men and women were not. In his introduction to *Leviathan*, Hobbes stated that, since life is mechanical, there's no reason that automatons can't be said to be alive. "For what is the *Heart* but a *Spring*," he said, "and the *Nerves*, but so many *Strings*, and the *Joynts*, but so many *Wheeles*, giving motion to the whole Body, such as was intended by the Artificer?"[13]

Hobbes went on to argue that a powerful government authority, the leviathan of the book's title, is required to restrain the savage impulses of human beings, lest the law of the jungle rule in society as it does in nature. What we have in a human being, Hobbes seemed to be suggesting, is a machine that could murder someone at any moment.

Darwinism accentuated the ambivalence of that conception dramatically. In 1863, four years after *On the Origin of Species* was published, Thomas Huxley, Darwin's leading advocate, wrote that the "question of questions" for humankind was "the ascertainment of the place which Man occupies in nature and of his relations to the universe of things." A year later he wrote an essay titled "On the Hypothesis That Animals Are Automata, and Its History," in which he affirmed the conclusion that "the living body is a mechanism." Unlike Descartes, Huxley included in that assessment human beings, with a significant qualification: we are, he said, "conscious automata." In other words, we're animals, hence mechanistic, but a higher order of animal, separated from the lower orders by our powers of reason and speech. Although there is "no absolute structural line of demarcation" between humans and animals, Huxley said, "at the same time no one is more strongly convinced than I am of the vastness of the gulf between civilized man and the brutes; or is more certain that whether *from* them or not, he is assuredly not *of* them."[14]

Darwin himself struggled mightily with these issues, with similarly mixed results, as the final paragraph of *The Descent of Man* attests. "Man may be excused for feeling some pride at having risen, though not through his own exertions, to the very summit of the organic scale," Darwin wrote. This achievement gives hope, he added, "for a still higher destiny in the distant future." Like Huxley, though, he added a qualifier, one that seemed to imply more than it said outright: "We must, however, acknowledge, as it seems to me, that man with all his noble qualities, with sympathy which feels for the most debased, with benevolence which extends not only to other men but to the humblest creature, with his god-like intellect which has penetrated into the movements and constitution of the solar system—with all these exalted powers—Man still bears in his bodily frame the indelible stamp of his lowly origin."[15]

I thought of the transhumanists when I read Darwin's comment that man, having risen to the summit of the organic scale, can hope for "a still higher destiny in the distant future." This as much as anything explains their hopes for the Singularity. If evolution has brought humans to their present stage of development, where do we go from here? Machines have the potential to move us past the biological impasse in which we're mired—and quickly. Thanks to exponential technological growth we can accelerate that pokey Darwinian timetable substantially.

Near the end of his life, Thomas Huxley seemed less certain that humans' place at the pinnacle of evolution was as secure as he'd earlier avowed, writing that man would never completely subdue the beast within. Here he was in agreement with the views of Freud, who was convinced, as Raymond Corbey put it, that an "impulsive ancestral apeman" roamed the depths of the human psyche. Whether Freud would have endorsed the fourth discontinuity as he did the first three is unknown, although in *Civilization and Its Discontents* he remarked that mankind had become "a prosthetic god." "When he puts on all his auxiliary organs," Freud said, "he is truly magnificent; but those organs have not grown on to him and they still give him much trouble at times."[16]

I bring all this up to underscore the odd position we're in today, when our place in nature and our relation to the universe of things, as Huxley put it, are more perplexing than ever. A word that seems to fit the current state of affairs is "ecotone." The *American Heritage Science Dictionary* defines an ecotone as "a transitional zone between two ecological communities." The definition adds that each of the two overlapping ecological communities in an ecotone retains its own characteristics in addition to sharing certain characteristics with the other community. That speaks to our confusion about where the line between humans, animals, and machines should be drawn. The etymology of "ecotone" implies as much: "tónos" is Greek for "tension."[17]

It used to be thought, and still is in some quarters, that when people stayed out in the wilderness they eventually devolved, steadily becoming more like the animals they lived with than the humans they'd left behind. Is it unreasonable to think that a transformation toward the other end of the continuum occurs when we surround ourselves with technology?

Human? Animal? Machine? Boundary negotiations are ongoing.

PART V
FEARLESS LEADERS

The age has an engine,
but no engineer.
RALPH WALDO EMERSON

CHAPTER 15

GAMBLERS

People are just curious. . . . What follows in
the wake of their discoveries is something
for the next generation to worry about.

WERNER VON BRAUN

Every so often in an era of technological enthusiasm, against the odds, a figure
emerges from the ranks of the technologists themselves who says, "Wait a min-
ute. Not so fast."

The best-known recent example is Bill Joy, the cofounder and former chief
science officer of Sun Microsystems, who in 2000 wrote a famous cover story in
Wired magazine titled "Why the Future Doesn't Need Us." Joy warned that rapid
developments in genetics, nanotechnology, and robotics threaten to escape
our control, posing unprecedented risks for the future of humankind. Failing
to recognize those risks, he said, the scientists and engineers now working so
feverishly to advance those technologies may one day find themselves in the
same position as the scientists and engineers of the Manhattan Project, who
recognized too late the fearful power they'd unleashed in the world.

Ever since the Industrial Revolution any number of Cassandras have tried to
alert us to the dangers hidden in the promises of technological advance. Sometimes their warnings have been heard, as Joy's were, but they have almost never
been heeded. Despite abundant evidence today that many of those warnings
were precisely on target, with few exceptions technological advance has continued, full throttle, regardless of consequences.

Another individual comes to mind who, a generation earlier, sounded
alarms much the same as Joy's. Like Joy, he was a respected member of the
scientific and technical communities, a position that bestowed upon his con-

cerns both credibility and conspicuous attention. Also like Joy, the attention soon faded and the advances he warned about proceeded apace.

I refer to Norbert Wiener, the father of cybernetics. I mentioned in my chapter on abstraction Wiener's role in the development of information theory and his "gloomy" speculations on where the technologies that emerged from that theory might take us. He took pleasure in playing the contrarian, but his concerns were genuine. "The world of the future will be an ever more demanding struggle against the limitations of our intelligence," Wiener wrote in 1964, "not a comfortable hammock in which we can lie down to be waited upon by our robot slaves."[1]

Wiener's groundbreaking work in automation technologies was one of the main reasons people had begun to think robot slaves would soon be handling our dirty work. He worried, though, that the rich would use automation to become richer at the expense of labor, turning human beings into robots. "Those who suffer from a power complex," he wrote in 1950, "find the mechanization of man a simple way to realize their ambitions." Sixty-five years later the likelihood that automation poses a serious and imminent threat to employment has become a topic of widespread comment and concern.[2]

Wiener was also far ahead of his time in recognizing that our habits of technology-fueled consumption pose significant threats to our health and to the environment. He predicted that within the foreseeable future we would be facing growing coal and gas shortages, growing scarcity of water with which to supply our cities, growing risks of infection due to increased air travel and antibiotic resistance, growing problems related to processed and synthetic foods, and growing risks of nuclear power accidents. A quote of his came to mind as I watched the news about the Fukushima meltdowns in Japan. "It is not necessary to bring in the consideration of war," he said, "for us to see how much more naked we lie to disaster than at any time before."[3]

Wiener did more than worry about these things. He spoke out about them, loudly and often, in best-selling books, in magazine articles, and in speeches. He also declined to participate in research projects—military and corporate projects, for the most part—that would make those threats more likely to materialize, a stance that cost him dearly in terms of his career, his pocketbook, and his reputation.

Wiener was hardly a saint. His writing reveals at times an unappealing pridefulness, not surprising in a man celebrated from a young age as a genius (his father promoted him relentlessly as a child prodigy) but nonetheless ironic, given his indignation at the arrogance of scientists. And although Wiener's depressions and personal eccentricities are understandable, given his upbringing, there's no question they played a role in the disappointments that plagued his career.[4]

Another irony is the fact that so many theorists today proclaim with utter certainty the universal applicability of information theory, given that Wiener, one of its founding fathers, harbored a lifelong distrust of certainty. I find it astonishing, and proof that Wiener probably *was* a genius, that at the age of ten he wrote a paper titled "The Theory of Ignorance." In it he dismissed those who presume that human knowledge is limitless and insisted that in the end it's impossible to know *anything* with absolute certainty. This is a fact of human existence, he wrote, that has been "far too often disregarded."[5]

Claims about human progress notwithstanding, that comment is as true today as it was then. If there is a single lesson, one key understanding, that I could drum into the mind of every technician on the planet, it would be the certainty of uncertainty. For despite their willingness to acknowledge uncertainty on the micro level and to use it to improve performance, technophiles consistently evince a depressingly broad degree of myopia in regard to uncertainty on the macro level. In other words, scientists and engineers will focus intently on the inconsistencies that appear within their specific projects and work diligently to get rid of them. At the same time they'll be perfectly willing to overlook the unpredictable results of their projects' interactions with other, supposedly unrelated technologies in the world at large. In doing so they ignore two fundamental principles:

1. There are no unrelated technologies.
2. The more powerful a given technology, the more widely its effects will radiate outward, the more difficult it will be to predict those effects, and the more damaging those effects can potentially be.

I said that scientists and engineers will "overlook" the unpredictability of the technologies they employ, which leaves open the possibility they may be unaware of that unpredictability. In truth I suspect it's more likely that they *are* aware of that unpredictability but choose to ignore it. Wiener was among many commentators who noted this characteristic and described it as suggesting a "gambler" mentality among scientists and engineers. "Gadget worshipers" was Wiener's name for those who are all too eager to employ the "magic" of the technologies they unleash without bothering to contemplate their ability to control that magic. "Of the devoted priests of power," he wrote, "there are many who regard with impatience the limitations of mankind, and in particular the limitation consisting in man's undependability and unpredictability."[6]

One reason unpredictability is a given in the technological project is that science and technology proceed by experiment, and experiment by definition entails a degree of risk. We can't know the outcome until we try it. As Rob Carlson, a leading advocate of synthetic biology, put it, "We generally cannot know whether a technology is, on balance, either valuable or beneficial until it is tested by actual use."[7]

Some of our giddier enthusiasts have taken that principle and run with it, insisting that there's nothing so uncool as caution. As Kevin Kelly put it in his book *Out of Control,* "To be a god, at least to be a creative one, one must relinquish control and embrace uncertainty. Absolute control is absolutely boring." Ray Kurzweil expressed similar sentiments in *The Singularity Is Near.* A necessary component in the forward motion of creation, he said, is "chaos," which provides "the variability to permit an evolutionary process to discover more powerful and efficient solutions."[8]

It's a fair point that a certain degree of risk is inherent in creative endeavor. (As noted in chapter 7, once a technological system is in place, control becomes a central concern.) When you're dealing with technologies that can potentially affect the well-being of millions of people, though, the scientists and technicians conducting the experiments are volunteering the rest of us to share the risk and to share the cost of the damage if things don't go as planned. For that reason an appropriate balance must be struck between the degree of risk taken and the degree of caution observed. At this point it's not at all clear that such a balance is being maintained.

One of the chief reasons Kurzweil is convinced the Singularity will materialize, for example, is the promise of nanotechnology, which he's counting on to heal virtually any sickness, eliminate pollution, and provide an endless supply of consumer goods. His anticipation of these wonders overlooks the fact that the biological and ecological effects of nanotechnology are, at this point, mostly unknown. Commercial development proceeds apace nonetheless, virtually unregulated.[9]

The same goes for synthetic biology, which enables, through the assembly of DNA sequences, the construction of life forms, including those that don't exist in nature. Enthusiasts in that field are actively cultivating a DIY approach, encouraging "bio-hackers" to invent and manufacture organisms, literally in their garages. Here, too, development proceeds with virtually no regulation, despite the fact that nobody knows how these organisms might behave when they're let loose in the environment, and despite the fact that the malicious release of a homemade virus could conceivably result in hundreds of thousands, if not millions, of deaths. Between 2008 and 2014, various agencies of the U.S. government spent $820 million on research into synthetic biology. Of that amount, less than 1 percent went to studying the technology's risks.[10]

It would be unfair to suggest that the advocates of nanotechnology and synthetic biology are ignoring the dangers. They're not. Conferences have been held to discuss safety measures that could be put in place; committees have produced position papers; guidelines have been proposed. It's also true that many of those pursuing these technologies do so idealistically, in hopes they might solve some of the world's most pressing problems. Nonetheless, there's no question that in the process of solving those problems, many also are hoping to become rich and famous. The mixture of altruism and ambition is intoxicating and not conducive to restraint. In the rush to exploit these world-changing technologies, restraint is often seen as a liability. There are also occasions when those who develop a technology would be unable to control its forward momentum even if they wanted to.

An especially powerful example is a gene-editing technique called CRISPR, which makes it easier, faster, and cheaper than ever before to alter the DNA of organisms, including humans. The promise of CRISPR (for "clustered regularly-interspaced short palindromic repeats") is that it can be used to repair the genetic defects that cause inherited diseases such as sickle cell disease, Huntington's disease, and certain forms of breast cancer—even, perhaps, Alzheimer's. Traits deemed undesirable in plants and animals can also be erased. From prevention it's a short step to enhancement: CRISPR makes it possible to add as well as remove genetic traits. Crops could be engineered to resist parasites; animals could be engineered to resist infections. One experiment created a litter of beagle puppies whose DNA had been manipulated to make them more muscular; another created a breed of hornless dairy cows. It may also be possible to engineer human babies to make them smarter, taller, better looking, or more athletic. This last prospect especially raises the question, Do we really want to go there?[11]

CRISPR is disquieting for another reason. Scientists can use it to create, with far greater reliability than had previously been possible, genetic characteristics that are inheritable, not only by individual descendants but also by the children of those descendants—indeed, by entire species. That's playing with the planetary gene pool on a whole different level. A team of scientists has tried to design a species of mosquito that could only produce male offspring, for example, thereby substantially reducing, if not eliminating altogether, the primary carrier of malaria. The problem is that, once a gene line has been altered, no one knows what ancillary effects organisms inheriting that alteration might have when they're released into the wild. This is especially true because evolution will go to work on the mutated genes we've introduced, producing new mutations with entirely unpredictable results.[12]

Because CRISPR is relatively easy and cheap to use, it has spread with amazing speed. Its ability to edit specific DNA sequences was first demonstrated in 2012; by 2014 more than 150 CRISPR-related patent applications had been filed and hundreds of millions of dollars invested in hopes of reaping scientific breakthroughs, Nobel prizes, and breathtaking profits. Scientists around the world have been using that money to manipulate the genetic makeup of dozens of organisms, among them pigs, rabbits, yeast, wheat, rice, fruit flies, and, of course, mice.[13]

The rapidity of this proliferation and the lack of authoritative oversight to ensure that experiments are being conducted safely began to concern even CRISPR's inventors. In July 2014 leading scientists in the field simultaneously published editorials in two journals urging researchers to proceed with caution and regulators to take action so that safety and ethical standards could be considered, adopted, and enforced. "Protect Society from Our Inventions, Say Genome-Editing Scientists" read the headline in *MIT Technology Review*. The urgency of the matter increased significantly less than a year later, when scientists in China announced they'd used CRISPR to edit the genes of human embryos. The embryos they manipulated wouldn't have been able to create a viable pregnancy, and the experiment failed, but the prospect of made-to-order babies, long a staple of science fiction, suddenly seemed very real, and very troubling. A line had been crossed, and the scientific community felt it could afford to wait no longer.[14]

In December 2015 several hundred scientists and ethicists from around the world convened in Washington, D.C., determined to recommend at least some preliminary guidelines. A position statement was issued declaring that the production of a human child using the CRISPR technique would be "irresponsible" until the safety of such a procedure could be assured and a societal consensus on its ethical use had formed. The statement did not call for a moratorium on all CRISPR research, however. To the contrary, continuing laboratory experiment was encouraged so that gene editing's potential benefits can be explored. Those experiments would include the editing of human cells, as long as no one tries to actually produce a child with them. Most of this research will be conducted by scientists, companies, and institutions that have financial interests in its outcomes.[15]

The statement isn't binding, and it's up to individual nations to decide what scientists within their borders will be permitted to do. In the meantime, it seems likely that the voluntary strictures on human experiments will be honored, though for how long is hard to say. Researchers and their sponsors will find it difficult to resist the temptation to exploit CRISPR's incredible powers. They may find it especially difficult to resist the pressure they're already receiving from individuals who have the diseases that gene editing might prevent. Antonio Regalado of *MIT Technology Review* interviewed several such people who desperately hoped CRISPR meant they wouldn't have to pass the maladies they'd inherited on to their children and grandchildren. That sentiment was underscored by the sobbing woman who stood up during a question-and-answer session at the Washington convocation to tell the gathered scientists that an inherited disease had killed her six-day-old son. Frustrated by the cautionary statements she was hearing from the podium, she expressed, in no uncertain terms, her wish that the editing of defective human genes proceed without delay. "Just frickin' do it," she said.[16]

Demands such as these suggest we'll be seeing the arrival of Huxley's brave new world sooner rather than later. CRISPR-altered animals and plants will likely be released into the wild sooner than that, with results that are both uncertain and permanent. Meanwhile, the possibility that CRISPR might fall into the hands of terrorists led a group of American intelligence organizations, including the CIA and the National Security Agency, to add genome editing to their list of potential weapons of mass destruction. Given CRISPR's "broad distribution, low cost, and accelerated pace of development," the group's report said, "its deliberate or unintentional misuse might lead to far-reaching economic and national security implications."[17]

That powerful technologies, once unleashed, are difficult if not impossible to control was one of the central concerns Bill Joy expressed in his famous piece for *Wired*, again using the Manhattan Project as an example. "The experiences of the atomic scientists clearly show the need to take personal responsibility, the danger that things will move too fast, and the way in which a process can take on a life of its own," he said. "We can, as they did, create insurmountable problems in almost no time flat. We must do more thinking up front if we are not to be similarly surprised and shocked by the consequences of our inventions. . . . Can we doubt that knowledge has become a weapon we wield against ourselves?"[18]

This seems roughly equivalent to something Norbert Wiener wrote in 1950: "Our fathers have tasted the tree of knowledge, and even though its fruit is bitter in our mouths, the angel with the flaming sword stands behind us."[19]

CHAPTER 16

CONSEQUENCES

> Though but a moment's consideration will teach,
> however baby man may brag of his science and skill,
> and however much, in a flattering future, that science
> and skill may augment; yet for ever and for ever, to the
> crack of doom, the sea will insult and murder him, and
> pulverize the stateliest, stiffest frigate he can make.
>
> HERMAN MELVILLE, *MOBY DICK*

This book went through several rewrites over a period of several years, and each time I revisited this chapter the news media were filled with stories that colorfully illustrated its theme.

During the first go-round, in the spring of 2010, daily reports updated how many millions of gallons of oil were pouring into the Gulf of Mexico as a result of the Deepwater Horizon disaster. At the same time Google and Facebook were acknowledging programming errors that had violated the privacy of millions of users and the Dow Jones industrial average mysteriously plunged nearly a thousand points in a matter of minutes, a chain-reaction "flash crash" later blamed on computerized trading programs. A year later, as I was working on my first revision, a devastating tsunami struck Japan, causing a series of meltdowns in the Fukushima nuclear plants that spewed radiation into the atmosphere, contaminating some eight thousand square miles of towns, villages, forests, and farmland.

Since then technology snafus of varying degrees of severity have continued to emerge with a regularity that should be surprising but isn't. Whether their frequency has increased or if it only seems that way is hard to tell, but it stands to reason that as our world becomes increasingly technicized, our technological misadventures will inevitably mount.

Certainly our reliance on digital communications has added hugely to the list. It no longer seems unusual to hear that a computer malfunction in the airline industry has stranded thousands of travelers or that hackers have stolen the personal data of millions of consumers. Add to these the officially sanctioned breaches of privacy revealed by Edward Snowden, who informed us that the National Security Agency had been secretly collecting the phone and Internet records of, it seemed, pretty much everybody.

Not all our technology problems are digital—far from it. Faulty ignition switches led to the recall in 2014 of 30 million GM cars and 124 deaths, while faulty airbags in a variety of makes and models led to the recall of at least 70 million cars, 11 deaths, and hundreds of injuries. A growing number of freight-train derailments have caused millions of gallons of oil to explode, setting entire neighborhoods ablaze, while chemical spills and pipeline ruptures continue to pollute beachfronts and waterways across the land. To cite one example, in January 2014 some 10,000 gallons of industrial chemicals leaked into the Elk River near Charleston, West Virginia, leaving 300,000 local residents without water to drink, cook with, or bathe in. According to the *New York Times*, this was the third major chemical accident in five years in a region known as Chemical Valley. Those affected were undoubtedly happy that thousands of cases of bottled water were trucked in, presumably not concerned for the moment that their discarded containers might end up in the Great Pacific garbage patch, a whirling mass of debris and sludge covering a million or so square miles in the northern Pacific.[1]

All these events demonstrate how the effects of powerful technologies radiate outward, producing in the process consequences that are both unintended and unexpected, often at velocities that exceed our ability to stop or contain them. This was another area in which Norbert Wiener was ahead of his time. In 1960 he wrote an essay for *Science* magazine titled "Some Moral and Technical Consequences of Automation." People are often under the misimpression, he said, that machines can do only what human beings tell them to do. This leads them to assume that machines are under our control. That might have been true once, he said, but not anymore: modern machines move too fast. "By the very slowness of our human actions," Wiener concluded, "our effective control of our machines may be nullified. By the time we are able to react to information conveyed by our senses and stop the car we are driving, it may already have run head on into a wall."[2]

The Wall Street flash crash showed just how prescient this remark had been. "We have a market that responds in milliseconds," said a finance expert quoted by the *New York Times*, "but the humans monitoring respond in minutes, and unfortunately billions of dollars of damage can occur in the meantime." A similar comment followed a shutdown of all United Airlines' flights due to a computer malfunction in the summer of 2015. "These are incredibly complicated systems," a security consultant told the *Los Angeles Times*. "There are lots and lots of failure modes that are not thoroughly understood. Because the systems act so quickly, you have this really increased potential for cascading failures."[3]

Events that are immediately disruptive get our attention, but suddenness and speed aren't necessary components of technological mishap. Edward Tenner, author of *Why Things Bite Back: Technology and the Revenge of Unintended Consequences*, points out that the amount of oil spilled by the *Exxon Valdez* pales in comparison to the amount of oil slowly leaking, day in and day out, from any number of underground storage tanks. The philosopher of technology Langdon Winner adds that unintended consequences can be direct, such as the poisoning of fish in a stream by pollutants, or distant and diffuse, such as the social and political impacts of television. The gradually accumulating effects of global warming have made it easier to deny or ignore its existence.

Unintended consequences aren't limited to large-scale events, either. To the contrary, technological power goes awry in countless ways in the hands of individual citizens, as any excursion on the nation's highways will affirm. Murphy's Law can always enter the equation, too. One day a bird dropped a piece of bread into the works of the massive Large Hadron Super Collider near Geneva, Switzerland, causing significant overheating of the world's most advanced and expensive particle accelerator.[4]

It's hard to remember how often unintended consequences came up before Edward Tenner gave us a catchphrase to describe them. Certainly, as I say, we've become increasingly inured to hearing about them today. But while it's undoubtedly true that the number and impact of such events are increasing, it's also true that unintended consequences are anything but a new phenomenon. A 1936 paper by the sociologist Robert K. Merton began by stating, "In some one of its numerous forms, the problem of unanticipated consequences of purposive action have been treated by virtually every substantial contributor to the long history of social thought."[5]

In his book *Autonomous Technology*, Langdon Winner offers a tour de force analysis of that history. So often have philosophers commented on the phenomenon of unintended consequences, Winner says, that it constitutes a "countertradition to the dream of mastery" in the history of Western thought. In this countertradition, "the world is not something that can be manipulated or managed with any great assurance. The urge to control must inevitably meet with frustration or defeat."[6]

Among those Winner cites as representatives of the countertradition to the dream of mastery are Machiavelli and Hannah Arendt. We think of Machiavelli as a designer of intricate schemes aimed at the successful control of events, and he was. He also recognized, however, the exigencies of *fortuna*, which he compared (in language eerily evocative of a tsunami) to "an impetuous river that when turbulent, inundates the plains, casts down trees and buildings, removes earth from this side and places it on the other; every one flees before it, and everything yields to its fury without being able to oppose it."[7]

Arendt's focus centered less on the uncertainties of fate and more on the individual's ability to topple the first domino. "Action," she said, "no matter what its specific content, always establishes relationships and therefore has an inherent tendency to force open all limitations and cut across all boundaries. . . . The reason why we are never able to foretell with certainty the outcome and end of any action is simply that action has no end. The process of a single deed can quite literally endure through time until mankind itself has come to an end."[8]

The fact that a consequence is unintended doesn't necessarily mean it's unforeseeable, or innocent. Langdon Winner makes this point, too. The intention of an industrial company that routinely dumps poisonous waste into a river is to get rid of the waste. The fact that this poisonous waste also kills the fish in the river is, strictly speaking, an unintended consequence, but hardly a surprising one. Unintended consequences shouldn't be construed as a convenient rationalization for evasions of responsibility.

Chances are that those who try *not* to evade responsibility won't be taken seriously. Norbert Wiener wasn't. Between the years of 1968 and 1972, the historian and physicist Steve J. Heims asked "a considerable number" of scientists and mathematicians what they thought of Wiener's social concerns and his preoccupation with the uses of technology. "The typical answer went something like this," Heims said: "Wiener was a great mathematician, but he was also an eccentric. When he began talking about society and [the] responsibility of scientists, a topic outside of his area of expertise, well, I just couldn't take him seriously."[9]

Heims adds that this dismissal on grounds of expertise wasn't surprising, given that taking seriously the issues Wiener raised "would demand reexamination of a whole set of received ideas." Not a few scientists and technicians had the same response to Bill Joy's article in *Wired*.[10]

Though I understand that the constraints of conformity operate as decisively in the science and high-tech communities as any other, I find the reluctance of those within those communities to more seriously consider the potential repercussions of their work surprising. These are not stupid people. What accounts for their willingness to look the other way?

The gambler mentality and the addictive attraction of wielding the power that science and technology bestow have a lot to do with it. That was how Wiener explained the drive of his fellow scientists to develop ever more powerful atomic bombs, even after they'd witnessed the devastation of Hiroshima and

Nagasaki. "Behind all this," he said, "I sensed the desires of the gadgeteer to see the wheels go round."[11]

Another explanation frequently offered is that work in these fields requires and rewards an intensity and dedication that shuts out anything but the task at hand. "Failing to understand the consequences of our inventions while we are in the rapture of discovery and innovation seems to be a common fault of scientists and technologists," Bill Joy said.[12]

The biologist E. O. Wilson identified the same problem in considerably harsher terms. "The vast majority of scientists have never been more than journeymen prospectors," he said.

> They are professionally focused; their education does not open them to the wide contours of the world. They acquire the training they need to travel to the frontier and make discoveries of their own—and make them as fast as possible, because life at the edge is expensive and chancy. The most productive scientists, installed in million-dollar laboratories, have no time to think about the big picture, and see little profit in it. The rosette of the U.S. National Academy of Sciences, which the 2,100 elected members wear on their lapels as a mark of achievement, contains a center of scientific gold surrounded by the purple of natural philosophy. The eyes of most leading scientists, alas, are fixed on the gold.[13]

Certainly there are exceptions to this rule. Still, the logic on display in the pronouncements of those most responsible for developing and exploiting the products of science and technology is often disappointing, to say the least. The exponents of synthetic biology are as good examples as any.

One of the leading independent companies in the field is called Amyris. Based in Emeryville, California, Amyris received a $42.5 million grant from the Bill and Melinda Gates Foundation to develop a vaccine against malaria. At the time of this writing, Amyris's chief executive officer was John Melo, who joined the company after spending many years as an executive at British Petroleum (now BP). Melo told Michael Specter of the *New Yorker* he understands why some people are "anxious" about synthetic biology. "Anything so powerful and new is troubling," he said. "But I don't think the answer to the future is to race into the past."[14]

This is an example of deflecting criticism by erecting a straw man. Other than the Unabomber, few critics have suggested that a "race into the past" is the only way to responsibly address the dangers posed by synthetic biology or any number of other new technologies. To suggest otherwise is to cast anyone who expresses misgivings into the role of Luddite, a rhetorical trapdoor often used to dispose of critics without actually engaging their concerns.

The *New Yorker* article included insights of similar depth from Amyris's chief biologist at the time, Jay Keasling. (He subsequently moved on to become chief executive officer of the Joint BioEnergy Institute, a government-funded research center dedicated to the development of synthetic alternatives to gasoline.) Keasling said he was baffled that anyone could question the pursuit of a technology that has the potential to save millions of lives. That he could be baffled is in itself baffling, given that it's a technology that also has the potential to *end* millions of lives, by, for example, advertently or inadvertently creating and releasing an infectious virus against which human beings have no immunity.

Drew Endy, the assistant professor of bioengineering at Stanford University, was more forthcoming, acknowledging to Specter that the threats posed by synthetic biology are "scary as hell." He was also among those who, in subsequent testimony before a presidential commission on bioethics, called for government regulation of the field so that those threats won't become realities.

Still, Endy doesn't see the potential for global catastrophe as sufficient reason not to proceed. Applying an engineer's logic to the problem, he told Specter that someone who builds a bridge that collapses may be prevented from building bridges again, but that doesn't mean nobody will ever be permitted to build a bridge again. We have no problem accepting risks in that arena, he said. The analogy is absurd. Few bridge collapses have the potential of wiping out hundreds of thousands, perhaps millions, of people. Nor do unsafe bridges self-replicate. Endy acknowledged as much in his testimony before the presidential commission. "I don't have any experience engineering machines that replicate," he said. "No engineers really do."[15]

The idea that it's time to take an engineering approach to biology is Endy's favorite theme. He often says he isn't interested in abstraction, that he's a nuts and bolts guy who likes to get things done. He's enthusiastic about wresting synthetic biology out of the hands of "the usual suspects," meaning guys wearing white coats in laboratories. In one interview he talked about being invited to speak at an upcoming Chaos Communication Congress, which he described as "the largest hacker meeting in Europe, about 4,000 people." He felt an affinity for the hackers. They are, he said, "People who like to make stuff, people who like to understand how things work." This interest naturally extends, Endy added, to genetic engineering. "They're very interested in learning how to program DNA."[16]

Endy isn't concerned, apparently, that hackers haven't exactly proved themselves faultless guardians of the public trust in the world of the Internet. He and other leaders of the field cite the hacker model to underscore their determination to push synthetic biology forward as quickly and as freely as possible by teaching as many people as possible to use it.

Endy is a model of restraint compared to the renowned physicist Freeman Dyson, whose article on synthetic biology for the *New York Review of Books* qualifies as a classic in the annals of technology enthusiasm. Titled "Our Biotech Future," it described in ecstatic terms the dawning era of "domesticated" biotechnology. The day will soon arrive, Dyson wrote, when Mom and Dad, Brother and Sis will be able to manufacture new species of plants and animals for fun and profit. Synthetic bio kits will give gardeners the chance to genetically engineer new varieties of roses and orchids; pet lovers will be able to breed new species of cats, dogs, parrots, and snakes, not to mention entirely new varieties of pets no one's ever seen before.[17]

Like Drew Endy, Dyson believes we'll be better off when the ability to manipulate DNA isn't the exclusive province of corporations like Monsanto, Genentech, and Exxon but is available to everyone, much as access to computers has now been liberated from the clutches of IBM and the Department of Defense. Once domesticated biology "gets into the hands of housewives and children," he wrote, we'll see "an explosion" in the diversity of living creatures.

> Designing genomes will be a personal thing, a new art form as creative as painting or sculpture. Few of the new creations will be masterpieces, but a great many will bring joy to their creators and variety to our fauna and flora. The final step in the domestication of biotechnology will be biotech games, designed like computer games for children down to kindergarten age but played with real eggs and seeds rather than with images on a screen. Playing such games, kids will acquire an intimate feeling for the organisms that they are growing. The winner could be the kid whose seed grows the prickliest cactus, or the kid whose egg hatches the cutest dinosaur.[18]

Dyson acknowledged that exploitation of this powerful technology entails risks. "Rules and regulations will be needed to make sure that our kids do not endanger themselves and others," he said. "The dangers of biotechnology are real and serious." He then offered a series of questions that need to be answered regarding the open availability of synthetic biology: Can it be stopped? Should it be stopped? If not, what limits should be placed on it? How should we decide what those limits would be, and how would the limits be enforced, nationally and internationally? "I do not attempt to answer these questions here," he concluded. "I leave it to our children and grandchildren to supply the answers."[19]

Dyson thus advocates taking the same approach to synthetic biology we've taken to virtually every other technological advance: Charge ahead, giddily and greedily; worry about the consequences later. As troublesome as they might be in the long run, however, most other technological advances don't pose the sorts of immediate catastrophic risks of synthetic biology. From Dutch elm disease and the zebra mussel to the gypsy moth and the Asian tiger mosquito, introductions of exotic species into unfamiliar ecosystems have created extensive environmental harm, including direct harm to human health. It's anybody's guess what will happen if kids and adults around the world begin creating and releasing organisms that are exotic in *every* ecosystem.[20]

Dyson's article reminds me of another point Norbert Wiener made in his *Science* magazine article. Speed is a relative quality, he said. Whether they move in milliseconds or advance over decades, technologies are by definition moving too quickly if we implement them before we've thought through and prepared for their potential effects. Powerful technologies impose on us the responsibility to take "an imaginative forward glance," Wiener said, acknowledging that such an exercise is "difficult, exacting, and only limitedly achievable." It's fairly obvious that leaving it to our children to sort out the answers fails to meet that standard.

Dyson's confidence that our children will be able to sort out the answers rests on a key technophilic assumption: that technology has the power to save us from technology. The technological fix for the technological problem underlies everything from computer programs that manage email overload to geoengineering schemes to counteract global warming. For the technophile, the answer to virtually every technological problem is more technology.

A case in point is the "Ecomodernist Manifesto," an ambitious proposal published in April 2015 by an influential group of environmentalists, academics, authors, and think-tank fellows, among them Stewart Brand. The manifesto is a blueprint—long on conceptual projection, short on specific detail—for "decoupling" humankind from nature, a process that the authors contend would permit both to thrive. Vigorous technological development would reconcile the seemingly antithetical goals of economic growth and preservation of the environment; technological efficiency would ensure success. In an ecomodernist future, human populations will be increasingly concentrated in efficient cities, fed by increasingly efficient agricultural techniques that deliver more food while consuming fewer natural resources. Poverty will be eradicated, while nuclear power delivers abundant energy for all.[21]

The ecomodernists sometimes call themselves ecopragmatists, a self-description intended to emphasize their philosophical distance from environmentalists who believe that, when it comes to technology, less is more. They seem not to have noticed that their pragmatic manifesto is as utopian as they come. To them the commencement of the Anthropocene—the name often used to designate the current geological era, when human activity shapes planetary conditions—represents a glorious opportunity to be embraced, rather than a presumptuous imposition on the natural order. But while the ecomodernists clearly believe they've presented an original, cutting-edge proposal, in fact they've essentially added an ecological gloss to the gospel of technological progress. In so doing they ignore the fact that for more than two centuries now the fundamental characteristics of technological development have consistently moved us in precisely the opposite direction.

As annoying as I find the managerial assurance of the ecomodernists, I'm realistic enough to concede that science and technology may well be our only hopes at this point, even if that means we've placed our fate in the hands of gamblers, and that the game will never end. In truth, technology doesn't fix technology, technology *demands* technology. Given that we seem unable to make even minor sacrifices of consumption and convenience, we probably have no choice but to stay, in some fashion, the technological course—not triumphantly, as the ecomodernists would have it, but with an attitude of triage that recognizes a state of "supreme emergency." The societies we've constructed are so utterly dependent on our machines that any attempt to abruptly disconnect would be spectacularly, fatally disruptive. Unless and until we find a way to reposition ourselves in relation to nature, we're pretty much stuck.[22]

This is what Langdon Winner meant by "the technological imperative." It's also what Norbert Wiener meant when he said that the angel with the flaming sword stands behind us.

CHAPTER 17

HUBRIS

The essence of great tragedy, Eric Bentley says, is the
realization by the self that it is totally unequipped to
confront the universe. We might venture that even more
tragic than any classical or Shakespearean drama is the
crisis of illumination when man realizes, much too late
for any last-minute panaceas, that he is unequal to the
task of dealing with a universe of his own manufacture.

PERRY MILLER

It's a truism that power corrupts, and at its most fundamental level technology is about power. It follows that arrogance and overconfidence may be natural by-products of technological power. That's a combination to be concerned about, given that the power of technology and the people who wield it can so decisively influence the entire spectrum of human affairs.

Another truism, one we've heard ad infinitum in the information age, is Francis Bacon's aphorism that knowledge is power. Less often do you hear Bacon's warnings about the dangers of hubris. "And it hath been noted," he wrote, "that those that ascribe openly too much to their own wisdom and policy, end infortunate."[1]

Bacon also famously warned that nature can't be conquered but must be obeyed, a maxim that seems hopelessly antiquated in an era when his heirs proudly proclaim that the very levers of life are in their hands.

Bacon reminded his readers that it was the devil who aimed to "ascend to be like the Highest," a comment that referred, according to the theologian Douglas Groothuis, to a passage from Isaiah:

You said in your heart,
  "I will ascend to the heavens;
I will raise my throne
above the stars of God; . . .
    I will ascend above the tops of the clouds;
    I will make myself like the Most High."[2]

The verse that precedes that passage makes it clear that the prophet was talking about pride coming before the fall:

How you have fallen from heaven,
    morning star, son of the dawn!
You have been cast down to the earth,
    you who once laid low the nations!

As I've mentioned, in a world increasingly tied to technology, unintended consequences regularly produce disruptions of various kinds, some annoying, others disastrous. Still, if previous experience is any indication, these missteps will neither slow the technological juggernaut nor unduly damage the self-confidence of its pilots. Tom Wolfe called the Wall Street buccaneers of the 1980s Masters of the Universe, but all they had was money. The masters of the universe today are Mark Zuckerberg, Larry Page, and Sergey Brin, quasi-mythical figures who possess not only fabulous wealth but the knowledge that they've literally changed the world. Bill Gates, meanwhile, has ascended to the rank of Master of the Universe, Emeritus. Steve Jobs has simply ascended.

When Apple introduced the iPad in 2010, *Time* magazine sent the British journalist and comedian Stephen Fry to profile Jobs. Fry opened his piece by confessing that in the course of his career he had met five British prime ministers, two American presidents, Nelson Mandela, Michael Jackson, and the queen, but that his interview with Jobs made him more nervous than any of those encounters. About the same time a flurry of stories appeared in the press that Jobs was occasionally replying personally to emails sent to him by regular folks. The lucky souls who had received one of these Jobsian missives reacted as if they had been touched by the divine.[3]

Aggression is an ally of hubris, if not its sibling. Herbert Marcuse described technology's essential motivation as totalitarian, which is a more direct way of describing the inherent aggression Heidegger described as "setting upon." So it is that a propensity for ruthless competitiveness is as manifest among the tech billionaires of Silicon Valley today as it was among the robber barons of the nineteenth century. Here, too, Steve Jobs serves as exemplar, given the degree to which he sacrificed, in both his professional and personal lives, ethical considerations in favor of success.[4]

The intrinsic relationship between technology and power is most obviously manifested in the prosecution of war, where advances in weaponry have proved decisive from the wielding of the first boulder to the guided missile. This brings to mind another founding father of the computer revolution, John von Neumann, whose enthusiastic pursuit of the tools of warfare made him in many ways Norbert Wiener's antithesis.

One of the twentieth century's premiere mathematicians and theorists, Neumann played a key role in the development not only of computers but also of the atomic bomb. While the devastation of Hiroshima and Nagasaki helped convince Wiener to withdraw from participation in military-related research, Neumann maintained close ties with the defense establishment and lobbied for the development of even more powerful nuclear weapons. At one point he argued for an early, preemptive strike against the Soviet Union. Neumann held similar views regarding technology as a whole; he was, according to historian Steve Heims, "deeply committed to unlimited technological progress, to be achieved with the greatest possible speed."[5]

Whereas Wiener was socially awkward and neurotic, Neumann was socially gregarious and supremely confident. Their relationship was marked by periods of intense collaboration and mutual discovery interrupted by equally intense disagreements, jealousy, and feelings of betrayal. One of the latter moments occurred in 1955, when Neumann wrote an essay on the future of technology that amounted to a public refutation of Wiener's widely publicized calls for technological restraint.

Published in *Fortune* magazine, it was an odd piece that, among other things, brushed aside an early recognition of global warming by predicting that technology will soon allow us to control the climate at will. Typical of enthusiasts, Neumann declared that the likelihood of unexpected consequences is no reason to curb progress; he was confident that any problem that arose could be successfully addressed. "Technologies are always constructive and beneficial, directly or indirectly," Neumann said. To restrain them would be "contrary to the whole ethos of the industrial age."[6]

Neumann was often so eager to exploit the potential of the technologies he was working on that he chafed at the relative conservatism of the military and corporate contractors who were paying for them. Whereas they wanted to balance development with various expediencies, such as profit or budgets, Neumann's interest was simply to push the project in question as far forward as the state of the art would allow. His eagerness to exploit nuclear weapons and to personally witness their tests may have led to his early death from bone cancer.[7]

In private communications Neumann expressed his belief that the sacrifice of human lives is an acceptable price to pay for technological progress. It's a sentiment that places him on a continuum that stretches, today, to the more radical proponents of transhumanism, who see the replacement of human beings by machines as something to be celebrated rather than feared.

Hans Moravec, for example, is a roboticist who predicts that humans will survive, if they survive at all, only by merging with machines. The writer Mark Dery once asked him whether such a future might prove dystopian for those unable to afford the upgrade. Concerns of that sort are "largely irrelevant," Moravec replied. "It doesn't matter what people do because they're going to be left behind, like the second stage of a rocket. Unhappy lives, horrible deaths, and failed projects have been part of the history of life on Earth ever since there was life; what really matters in the long run is what's left over. Does it really matter to you today that the tyrannosaur line of that species failed?"[8]

This is the sort of attitude Bertrand Russell had in mind when he spoke of the modern tendency toward "cosmic impiety." But again, as the Bible's story of the Tower of Babel suggests, some measure of impiety may be built into the technological project. For all his talk of humility, Bacon's ambitions for his scientific method can hardly be described as modest. His goal was the restoration of humankind to its prelapsarian state, after which we'd regain our rightful dominion over creation. At one point Bacon reminded his readers that they were "not animals on their hind legs, but mortal gods," a maxim echoed centuries later by Steward Brand's mission statement in the *Whole Earth Catalog*: "We are as gods and might as well get good at it."[9]

The godlike aura enjoyed by those who create or wield technologies is, of course, a direct reflection of the godlike power of the technologies themselves. The pyramids of ancient Egypt expressed and affirmed the power of the pharaohs who built them, just as the great cathedrals of Europe expressed and affirmed the powers of the Catholic Church. The pyramids and cathedrals are examples of what historian David E. Nye has called "the technological sublime." The phrase refers to the feelings of astonishment, wonder, and fear engendered by overwhelming displays of technological power. Edmund Burke's classic definition of the sublime describes it as that which produces "the strongest emotion which the mind is capable of feeling." For Burke, terror was a key element of the sublime experience because it provided the necessary dislocation from habitual thought. As Nye puts it, the sublime strikes us dumb with amazement; in its shadow we feel insignificant.[10]

For much of history we encountered the sublime through nature, but gradually technology has become its most frequent source. Indeed, the technological sublime often achieves its sublimity by "conquering" nature in some fashion, whether by overcoming tremendous obstacles to build a canal or by harnessing the power of the atom. Among examples of the technological sublime listed by Nye are Hoover Dam, the Golden Gate Bridge, the Empire State Building, the Statue of Liberty, Henry Ford's Highland Park and River Rouge automobile assembly plants, spectacular fireworks and light displays, and the thunderous roar of a rocket as it's launched into space. He specifically excludes warfare from his analysis, but certainly the bombings of Hiroshima and Nagasaki, the United States' "shock and awe" bombing of Baghdad in 2003, and the terrorist attacks of 9/11 would fit the definition as well.

As I say, the worshipful responses inspired by powerful technologies naturally extend to those who create or wield those technologies, with the result that they can become identified, in their minds as in others', as intermediaries between God and humankind. "Examine the endless varieties of machinery which man has created," read an 1831 essay in the *North American Review*. "Mark how all the complicated movements co-operated, in beautiful concert, to produce the desired result.... We accordingly believe there is nothing irreverent in the assertion, that the finite mind in no respect approximates so nearly to a resemblance of the Infinite Mind, as in the subjugation of matter, through the aid of Mechanism, to fixed and beneficial laws,—to laws ordained by God, but discovered and applied by man."[11]

What the *Review*'s essayist failed to recognize was that the transcendent experience of the technological sublime would come to rival that of traditional religion. Henry Adams saw that shift coming. "The Dynamo and the Virgin," the most famous passage in his autobiography, is a characteristically subtle and ironic meditation on that theme. It is also one of our most penetrating descriptions of the technological sublime.

Writing in the third person, Adams described his visit with a friend to the Paris Exposition of 1900. The friend, who had no interest in anything in the exhibition other than new applications of power, strode directly to the Gallery of Machines, the hall in which, as Adams put it, "the forces" were displayed. There, Adams's encounter with a massive generator of electricity—a dynamo—provoked an epiphany that epitomized the century to come.

To him [Adams's friend], the dynamo itself was but an ingenious channel for conveying somewhere the heat latent in a few tons of poor coal hidden in a dirty engine-house carefully kept out of sight, but to Adams the dynamo became a symbol of infinity. As he grew accustomed to the great gallery of machines, he began to feel the forty-foot dynamo as a moral force, much as the early Christians felt the Cross. The planet itself seemed less impressive, in its old-fashioned, deliberate, annual or daily revolution, than this huge wheel, revolving within an arm's length at some vertiginous speed, and barely murmuring—scarcely humming an audible warning to stand a hair's-breadth further for respect of power—while it would not wake the baby lying close against its frame. Before the end, one began to pray to it; inherited instinct taught the natural expression of man before silent and infinite force.[12]

Forty-five years later, J. Robert Oppenheimer would experience a similar sense of awe as he watched the first atomic bomb explode over the New Mexico desert. A verse from the Bhagavad Gita came to his mind: "I am become death, destroyer of worlds." Twenty-four years after that, Norman Mailer, in Florida to witness the launch of the Apollo 11 moon flight, had a somewhat less cerebral reaction as the roaring waves of sound from the blast-off washed over him: "Oh my God! oh my God! oh, my God! oh, my God!"[13]

Perhaps it was inevitable that people who were enthralled by early examples of the technological sublime began to wonder if the engineers who built them might be just as proficient at steering the ship of state. A new political philosophy emerged: technocracy. Simply put, technocracy holds that the only people who really know what they're doing are the experts—engineers in particular—and that we'd all be better off if they were running the show. One of the first technocrats was Henri Comte de Saint-Simon, who argued during the French Revolution that management of "the governmental machine" ought to be turned over to people who could properly administer it, namely, industrialists, scientists, and technicians. Saint-Simon's assistant, Auguste Comte, founded a Religion of Humanity, in which scientists and engineers would become the new priests.[14]

Technocracy enjoyed a period of particular influence in the United States in the early decades of the twentieth century, blossoming at one point into a full-fledged movement. Its influence was an early example of the idea that only more technology, or technological thinking, could solve the problems technology was creating. As the powers of industrialism became the dominant forces in society, and as the society those forces created became increasingly complex, people began to believe that politicians were no longer able to understand what was happening, much less control it. What was needed was efficiency and planning, which is what the technicians did best.[15]

Engineers were in demand and universities were rapidly adding or expanding degree programs to supply that demand. The number of American engineering graduates increased from 100 a year in 1870 to 4,300 a year in 1914. With those degrees came a newfound sense of expertise and authority. What had been a trade became a profession.[16]

The civil engineering graduates of Cornell University were urged to recognize and embrace that authority when they met for their first annual reunion in June 1905. The speaker for the occasion was the engineer, editor, and manufacturer Henry Goslee Prout, whose address was titled "Some Relations of the Engineer to Society."

"My proposition," Prout said, "is that the engineer more than all other men will guide humanity forward until we come to some other period of a different kind. On the engineer and on those who are making engineers rests a responsibility such as men have never before been called upon to face, for it is a peculiarity of the new epoch that we are conscious of it, that we know what we are doing, which was not true in either of the six preceding epochs, and we have upon us the responsibility of conscious knowledge."[17]

Two men especially influenced the emergence of the "myth of the engineer" in the early twentieth century. One was Frederick Winslow Taylor. If engineers could employ "scientific management" to make a factory run at peak efficiency, it was natural to ask why the same skills couldn't be applied to the operations of government. The other key voice was that of the economist and social critic Thorstein Veblen. Best known today as the man who coined the phrase "conspicuous consumption," Veblen relentlessly attacked the wastefulness of American business. Overproduction and overselling of useless goods were ruining the country, he argued. The solution was to turn policy and administration over to "skilled technologists" who would exercise "systematic control" over the economy.[18]

Somehow the ascent of the engineers that Veblen and others envisioned never materialized. Despite their growing professional confidence, they seemed personally reluctant to pursue broader political power. Veblen couldn't conceal his disdain. "By settled habit," he fumed, "the technicians, the engineers and the industrial experts, are a harmless and docile sort, well fed on the whole, and somewhat placidly content with the 'full dinner-pail,' which the lieutenants of the Vested Interests habitually allow them." This pattern of political passivity would be repeated by the bland conformity of the typical engineer in the 1950s and by the disinclination of most engineers to actively support the values of the counterculture movement of the 1960s.[19]

The idea that engineering could set the country right reemerged with a vengeance during the Depression. Citizens struggling to put food on the table were receptive to the idea that technology, properly managed, could deliver them in short order to an era of abundance and leisure. It was at this point that Technocracy, fed by prodigious publicity and prominent supporters in academia and the press, became a mass movement. Chapters in cities around the country boasted memberships that ran into the hundreds of thousands, though subsequent studies have shown that few engineers signed on. In addition to Technocracy newspapers and Technocracy study courses, there was a Technocracy uniform, a Technocracy medallion, and fleets of Technocracy automobiles and motorcycles. The color scheme was gray.

The movement's basic principles can be discerned from one of its educational pamphlets. "Technocracy's scientific approach to the social problem is unique, and its method is completely new," it read. "It speaks the language of science, and recognizes no authority but the facts. In Technocracy we see science banishing waste, unemployment, hunger, and insecurity of income forever.... We see science replacing an economy of scarcity with an era of abundance ... [and] we see functional competence displacing grotesque and wasteful incompetence, facts displacing guesswork, order displacing disorder, industrial planning displacing industrial chaos."[20]

The momentum of the Technocracy movement faltered when a series of newspaper reports exposed its most prominent leader, Howard Scott, as a fraud. He'd given himself the title of "Chief Engineer" but his engineering credentials turned out to be bogus, as were the "scientific" facts he used to justify his proposed reforms.

Other image problems emerged. One, according to historian William Atkin, was the widespread belief that engineering as a profession "seemed to produce undesirable qualities in its members," namely, that they assumed the characteristics of their machines. President Herbert Hoover, an engineer by profession and an apostle of scientific management, embodied those suspicions, being widely perceived as incapable of responding with human feelings to the problems of the people. Worse, he was ineffective. The deeper fear was that the engineers were conspiring to turn *everyone* into a machine. Technocracy's literature did little to counter that impression. One pamphlet stated that "the human being is an engine" and urged that the masses be "conditioned" in order to obtain "the highest possible percentage of proficient functional capacity."[21]

Such statements are evidence of a fatal contradiction at the movement's core: it hoped to gain mass support for an elitist agenda. If the technological society had become too complex for the politicians to understand, it was also over the head of the average citizen. This remains a core problem in the governance of the technological society to this day. Political bodies like the President's Commission on Bioethics consistently state that the public needs to be included in policy decisions regarding potentially dangerous technologies such as synthetic biology and nanotechnology, and researchers involved in developing those technologies agree. An admirable goal, perhaps, yet polls consistently suggest that the vast majority of Americans have never heard of those technologies. The philosophers of technology Langdon Winner and Witold Rybczynski have both pointed out that in contemporary democracies voters responsible for making informed decisions regarding advanced technologies when they have trouble remembering what they studied in their high school science classes. It's also clear from the proceedings of bodies like the Presidential Commission on Bioethics that the old technocratic complaint still applies: the politicians responsible for regulating those advanced technologies don't understand them, either. Thus we rely on the testimony of the experts. But the experts are biased, and how do we know which experts to believe?

For insight into the technocratic mind-set at work, it would be hard to find a better example than Google. The company is run, with an unusual degree of control for a corporation of its size, by three computer scientists, cofounders Larry Page and Sergey Brin and Executive Chairman Eric Schmidt. They are, as Google's vice president of global communications and public affairs put it, "ideological technologists." Larry Page, who in 2011 took over the role of chief operating officer from Schmidt, explained his management philosophy to the journalist Ken Auletta. "There is a pattern in companies," he said,

> even in technological companies, that the people who do the work—the engi-
> neers, the programmers, the foot soldiers, if you will—typically get rolled over
> by the management. Typically, the management isn't very technical. I think
> that's a very bad thing. If you're a programmer or an engineer or a computer
> scientist, you have someone tell you what to do who is really not very good at
> what you do, they tell you the wrong things. And you sort of end up building the
> wrong things; you end up kind of demoralized. You want a culture where the
> people who are doing the work, the scientists and engineers, are empowered.
> And [where] they are managed by people who deeply understand what they are
> doing. That's not typically the case.[22]

For Page, these were far more than idle sentiments. In 2001, when Google was in the midst of its meteoric rise from fresh-faced start-up to major inter-national player (Eric Schmidt would arrive as CEO later that year, specifically to add some adult ballast to the executive ranks), Page decided that a growing layer of managers at the company was stifling the creativity of engineers. His solution was simple: he fired all the managers. The internal reaction to this de-cision—from engineers as well as managers—was so negative that it was soon rescinded, but the conviction that managers had no business telling engineers what to do clearly lingered.[23]

Eight years later an internal survey at the company revealed, to the sur-prise of many, that employees overwhelmingly believed that the most effec-tive managers were those who could relate to people as well as technology. "In the Google context, we'd always believed that to be a manager, particularly on the engineering side, you need to be as deep or deeper a technical expert than the people who work for you," Laszlo Bock, Google's head of human re-sources, told the *New York Times*. "It turns out that that's absolutely the least important thing. It's important, but pales in comparison. Much more impor-tant is just making that connection and being accessible." In response, Google instituted an extensive new training program to teach managers how to better relate on a personal basis with their subordinates. Bock told the *Times* that the program had proved a resounding success, but the language he used suggested Google would remain technocratic to its core. "We were able to have a statistically significant improvement in manager quality for 75 percent of our worst-performing managers," he said.[24]

"Technocracy" is a word that's rarely heard in American political circles anymore, although, like other big-time players, major corporations built on technology have long maintained active and effective lobbying presences in Washington. During their early years the upstarts of Silicon Valley showed little interest in that game, and in many cases actively disdained it. As their power has grown, that has changed, dramatically.

Google, for example, expanded its Washington lobbying forces from a one-person office in 2005 to a "lobby corps" of more than one hundred by 2014, according to the *Washington Post*. In 2012 it spent more money on influencing federal policy than any other company save one, General Electric. It also spends considerable sums to influence policy in Europe, where it has enlisted the help of American senators and members of Congress (whose election campaigns have benefited from Google contributions) to lobby against a major antitrust suit filed by the European Union.[25]

Another example is FWD.us, a political action committee focused on liberalizing American immigration law, a key issue for companies hoping to import foreign technology talent. The public face of FWD.us is Mark Zuckerberg's. Other founding or contributing members include a Who's Who of Silicon Valley heavyweights, from Bill Gates of Microsoft and Reed Hastings of Netflix to Marissa Mayer of Yahoo and Brian Chesky, cofounder and CEO of Airbnb. "Tech is in our DNA," the FWD.us website declares. "Our founders, members and supporters come from the tech community. FWD.us provides a way for us to come together to create a potent new political movement. . . . We know that politics isn't pretty and that sometimes it can be tempting to opt out. But when it comes to fixing this country's biggest challenges, disengagement just isn't an option."[26]

Engagement doesn't automatically mean victory, of course. FWD.us's campaigns on immigration reform have stumbled badly, and another Zuckerberg initiative—his donation of $100 million to reform the Newark, New Jersey, public school system—seems to have gone nowhere. Still, at this point Silicon Valley has enough money at its disposal to win more battles than it loses. In September 2015 the economist and former secretary of labor Robert Reich wrote an op-ed piece for the *New York Times* arguing that the political influence of Google, Amazon, Microsoft, and Facebook had made them essentially immune from federal regulation. The article's headline bluntly summarized his point: "Big Tech Has Become Way Too Powerful."[27]

Silicon Valley's corporate elite clearly believe that their political activism, like their management styles, will be more enlightened than that of their predecessors. Whether that proves to be so in the long run remains to be seen. Certainly the historical record of engineers steering the ship of state isn't encouraging. Herbert Hoover was a mining engineer and advocate of the technocracy-influenced efficiency movement, which sought to propagate the "scientific management" principles of Fredrick Winslow Taylor. His administration and his reputation will forever be tarnished by the Great Depression; the stock market crashed nine months after he took office. The second most prominent episode of technocratic influence in American politics—the Vietnam War as prosecuted by then-secretary of defense Robert McNamara—is another of the nation's most painful memories.[28]

McNamara earned an economics degree from the University of California at Berkeley and an MBA from the Harvard Business School. Upon graduation he joined the accounting firm Price, Waterhouse, but after two years returned to Harvard as an assistant professor of business administration. When World War II started he helped organize the Statistical Control Office of the Army Air Force, where he collected and analyzed information on deployment of resources, including human resources. *Fortune* magazine called McNamara's methods a "super application of proved business methods to war."[29]

McNamara's work with the Statistical Control Office led to his hiring by the Ford Motor Company a few months after the war ended. Ford at the time was suffering from slumping sales and a management team in disarray. McNamara was brought in for his organizational and accounting skills; he knew nothing about making or selling cars. The same skills brought him to the attention of John F. Kennedy, who appointed McNamara head of the Department of Defense in 1961.

The Pentagon was as big a mess as Ford had been when McNamara took over. His goal, he said at the time, was "to bring efficiency to a $40 billion enterprise beset by jealousies and political pressures." His management style was relentlessly disciplined. Meetings were short and scheduled to the minute; reports were to be submitted on paper, accompanied by ample statistical documentation. Facts were the only currency McNamara honored; philosophical discussions were off limits. Lyndon Johnson compared him to "a jackhammer." "No human being can take what he takes," Johnson said. "He drives too hard. He is too perfect."[30]

A passage from a book McNamara published in 1968, *The Essence of Security*, describes his philosophy.

> Some critics today worry that our democratic, free societies are being overmanaged. I would argue that the opposite is true. As paradoxical as it may sound, the

real threat to democracy comes, not from overmanagement, but from under-management. To undermanage reality is not to keep free. It is simply to let some force other than reason shape reality. That force may be unbridled emotion; it may be greed; it may be aggressiveness; it may be hatred; it may be inertia; it may be anything other than reason. But whatever it is, if it is not reason that rules man, then man falls short of his potential.[31]

McNamara quickly took a leading role in determining American policy on Vietnam, overpowering Secretary of State Dean Rusk. In 1962 he returned from his first tour of the Asian theater brimming with confidence. "Every quantitative measurement we have shows we are winning this war," he said. His analysis led him to expect victory within three to four years.[32]

As we now know, that analysis was mistaken. McNamara himself began to doubt that the war was winnable long before he left the Pentagon in 1967, but he didn't say so publicly. Tens of thousands of Americans and hundreds of thousands of Vietnamese ultimately died. David Halberstam, who witnessed the disaster firsthand as a young reporter for the *New York Times*, summed up McNamara's stewardship of the war with bitter fury: "He knew nothing about Asia, about poverty, about people, about American domestic politics, but he knew a great deal about production technology and about exercising bureaucratic power. . . . He symbolized the idea that [the administration] could manage and control events, in an intelligent, rational way. Taking on a guerilla war was like buying a sick foreign company; you brought your systems to it."[33]

McNamara was haunted by the war for the rest of his life. He confessed his mistakes in a 1995 book and later in a documentary film by Errol Morris, *The Fog of War: Eleven Lessons from the Life of Robert S. McNamara*. Lesson no. 2: "Rationality will not save us."[34]

One thing that's clear from McNamara's confessions is that he always considered himself a moral man. He firmly believed, at the time, that his actions as secretary of defense were appropriate and honorable. This is yet another truism: the road to hell is paved with good intentions. Faust in Goethe's rendering goes to his death believing his vast development project, by driving back the ocean, will open "room to live for millions," disregarding the innocent souls whose land and lives have already been taken. "Goethe's point," wrote Marshall Berman, "is that the deepest horrors of Faustian development spring from its most honorable aims and its most authentic achievements."[35]

Unlike Faust, Mary Shelley's character ultimately recognizes the chaos he's created. In his student days, inspired by lectures that alert him to the "almost unlimited" powers science was putting at man's disposal, Frankenstein imagines he will "pioneer a new way, explore unknown powers, and unfold to the world the deepest mysteries of creation." On his deathbed he finds memories of those ambitions serve "only to plunge me lower in the dust."

> All my speculations and hopes are as nothing; and, like the archangel who as-
> pired to omnipotence, I am chained in an eternal hell. . . . I trod heaven in my
> thoughts, now exulting in my powers, now burning with the idea of their effect.
> From my infancy I was imbued with high hopes and a lofty ambition; but how
> am I sunk! Oh! My friend, if you had known me as I once was you would not rec-
> ognize me in this state of degradation. Despondency rarely visited my heart; a
> high destiny seemed to bear me on until I fell, never, never again to rise.[36]

All of this reminds me of Norbert Wiener's warning that "a sense of the tragic" is a prerequisite to the appropriate exercise of scientific and techno-logical power. Someone with an appreciation of how badly things can go wrong will understand, he said, that the only true security comes from "humility and restrained ambitions."[37]

I'm reminded, too, of another of Wiener's warnings, this one in the last book he wrote. Technology is a two-edged sword, he said, "and sooner or later it will cut you deep."[38]

CONCLUSION

There's a built-in expectation that the conclusion of a book like this one will wrap things up somehow. A trace of the technocratic mind-set flickers in that thinking: There's a fix for everything; just follow instructions! It's not that simple, of course. Nonetheless, I see no harm in mentioning two general suggestions that would, if widely and comprehensively pursued, move us in a positive direction.

The first of these is restraint. Cut back, on everything. It's remarkable how consistently this most obvious of all solutions is ignored, as if the idea that we could happily subsist with less is somehow unthinkable. The historian Perry Miller, referring to Thoreau's review of Etzler's *The Paradise within the Reach of All Men*, asked how many of those machines littering the ground we could choose to live without. "In this interrogation," he said, "consists the fundamental query for the mind of America."[1]

Useful primers for those wishing to pursue restraint as a way of life include Ivan Illich's *Tools for Conviviality*, E. F. Schumacher's *Small Is Beautiful*, and Bill McKibben's *Enough*. There are many others. The "degrowth" movement offers the single most hopeful path I can think of.

My second suggestion is so naive I'm embarrassed to mention it, but what the hell. I'd like to see us pay some attention to redressing the imbalance, in the culture in general and in education in particular, between means and ends. Every day we learn new ways to propel ourselves, but we still have trouble remembering where it was we were heading and why. This refers back to the classic/romantic split I discussed in part I. Thanks to the digital revolution, the gulf between the two orientations has widened steadily in recent years. As technology's march to dominance continues, we grow ever more lopsided.

Confronting that march is analogous to standing in front of a tank, which is why it feels naive to suggest it. In the face of the practical results science and technology have produced, romantic concerns regarding spiritual welfare and contemplative wisdom have always seemed softheaded. That doesn't mean they're wrong; it just means the felicities they promise are harder to put a finger on. In an age of technique, what can't be measured doesn't matter. Never mind. Challenge the materialist ethic wherever possible, for fun if not for profit.

Although the intent of this book has been to question our assumptions regarding the technological project, there are fundamental questions I still find hard to answer. One is whether the benefits of technology have made us better people, morally, than we were without them.

A tally of pros and cons offers, again, no simple answers. We no longer torture people to death in the public square, for example. Nor do we take slavery, rape, or starvation for granted, at least not as much as we used to. We have boxers but not gladiators; trial by television but not witch trials. All good, but far from the whole story. Thanks to our technologies we've found ways to incinerate more than a hundred thousand people almost instantly and to poison entire rivers with chemical waste. Also thanks to our technologies we now have the means at our disposal to destroy civilization as we know it—several means, in fact. The threat of global warming alone suggests that many of the gifts technology has bestowed may soon be eclipsed. When populations start fighting for food and shelter, count on whatever refinements we think we've acquired to disappear.[2]

A second fundamental dilemma needs to be acknowledged. I've mentioned that we may have no choice at this point but to rely on technology to solve the problems technology has created. Partly for that reason, it's also true that technology remains one of our greatest outlets for creative energy and optimistic enthusiasm, an engine of dreams as well as nightmares.

This has long been a key argument of the technophiles—I'm reminded of Stewart Brand's insistence that the possibility of establishing human colonies in space was worth entertaining if only because it would excite the imaginations of young people. Often these sorts of justification come across as childish and self-serving, but it can't be denied they contain more than a grain of truth.

A lot of people are using technology to pursue truly noble ambitions. With any luck some of them may be able to make significant contributions while avoiding the pitfalls in their way. The fact that the frontier myth has become a cautionary tale can't mean that every adventure of discovery has to be curtailed. The gospel of progress, as battered and discredited as it is, persists.

The question is whether we'll find a way to steer our dreams of progress in a sane direction. To the degree technology can assist us in achieving those ambitions, I'm all for it. Ideally we'll be able to make wise choices about which technologies to use and which technologies to forgo for our collective benefit. So far we haven't demonstrated an ability to make those distinctions, but you never know.

NOTES

Introduction

1. Borgmann, *Technology and the Character of Contemporary Life: A Philosophical Inquiry* (Chicago: University of Chicago Press, 1984), 208. The emphasis in the quotation is Borgmann's.

Chapter 1. The Paradise within the Reach of All Men

1. Ross Andersen, "Exodus," *Aeon*, September 30, 2014, http://aeon.co/magazine /technology/the-elon-musk-interview-on-mars/, accessed September 22, 2015.
2. On Andreessen, see Andrew Leonard, "Tech's Toxic Political Culture: The Stealth Libertarianism of Silicon Valley Bigwigs," *Salon*, June 6, 2014, http://tinyurl.com /nlt342x, accessed December 11, 2015.
3. Derek Jacoby, "George Church and the Potential of Synthetic Biology," *O'Reilly Radar*, November 9, 2012, http://oreil.ly/18F9siw, accessed February 2, 2013. Schmidt comments are from the Zeitgeist America conference, 2012, http://youtu.be/kUHF43xjMJM, accessed February 2, 2013.
4. Eric Anderson, cofounder and cochairman, Planetary Resources, webcast, April 24, 2012, http://bit.ly/1a2UoJp, accessed February 3, 2013.
5. Ford quoted in David E. Nye, *Henry Ford: Ignorant Idealist* (Port Washington, N.Y.: Kennikat Press, 1979), 71.
6. Eric Drexler, *Engines of Creation: The Coming Era of Nanotechnology* (New York: Anchor, 1987), 81. Note that Drexler's visions of nanotechnology's potential are not shared by other scientists in the field. See David Rotman, "Will the Real Nanotech Stand Up?," *MIT Technology Review*, March 1, 1999, http://tinyurl.com/hjgqv2u, accessed September 22, 2015.
7. Ray Kurzweil, *The Singularity Is Near: When Humans Transcend Biology* (New York: Penguin, 2005), 9.
8. Ibid., 303–7.
9. Ibid., 28–29, 318, emphasis added.
10. Ibid., 134, 29.

11. Enthusiast quotes: "The Spirit of the Times; or, The Fast Age," reprinted in *The United States Review: Democracy*, vol. 2, ed. D. W. Holly (New York: Lloyd and Brainard, 1854), 261; David E. Nye, *American Technological Sublime* (Cambridge, Mass.: MIT Press, 1999), 60; Merritt Roe Smith, "Technological Determinism in American Culture," in *Does Technology Drive History?: The Dilemma of Technological Determinism*, ed. Merritt Roe Smith and Leo Marx (Cambridge, Mass.: MIT Press, 1998), 5.

12. Howard P. Segal, *Technological Utopianism in American Culture* (Chicago: University of Chicago Press, 1985), 21.

13. Ibid.

14. There were at least three editions of *The Paradise within the Reach of All Men*, published, respectively, in 1833, 1836, and 1846. Quotations here are from the 1836 edition, published by John Brooks of London and available online at Google Books. British spellings have been altered to conform to American style. Biographical material on Etzler is from James M. Morris and Andrea L. Kross, *Historical Dictionary of Utopianism* (Lanham, Md.: Scarecrow Press, 2004).

15. Etzler, *Paradise within the Reach*, 1–2.

16. Ibid., 11–13.

17. Ibid., 47–48.

18. Thoreau's review is reprinted in *Changing Attitudes toward American Technology*, ed. Thomas Parke Hughes (New York: Harper and Row, 1975), 87–111. Also see Robin Linstromberg and James Ballowe, "Thoreau and Etzler: Alternative Views of Economic Reform," *American Studies* 11, no. 1 (Spring 1970): 20–29.

19. Thoreau, review in *Changing Attitudes toward American Technology*, 91.

20. Ibid., 109.

21. On harnessing the tides, see "New Research Supports the Huge Potential of Tidal Power," *Kurzweil Accelerating Intelligence News*, January 18, 2013, http://bit.ly/18Faon9, accessed February 3, 2013. On harnessing the wind, see Philip Warburg, *Harvest the Wind: America's Journey to Jobs, Energy Independence and Climate Stability* (Boston: Beacon Press, 2013).

Chapter 2. Absolute Confidence, More or Less

1. Robert Wright, "The Evolution of Despair," *Time*, August 28, 1995, available at http://tinyurl.com/p9ffqx7, accessed April 1, 2016; Kevles, *New Yorker*, August 14, 1995, 2.

2. Alston Chase, *Harvard and the Unabomber: The Education of an American Terrorist* (New York: Norton, 2003), 89–91.

3. Ibid., 89.

4. The National Science Board study is available at http://tinyurl.com/jkwsfbg. The Pew study is available at http://tinyurl.com/povkdrm. Both accessed November 26, 2015.

5. Reports on the studies cited can be found at http://bit.ly/18FaHoY; http://tinyurl.com/gwj3xv8; and http://bit.ly/18FaR8A. Accessed January 29, 2013, and April 20, 2013.

6. For Kaczynski's comment regarding the impossibility of separating good from bad technologies, see the Unabomber manifesto (actual title: "Industrial Society and Its Future"), paragraph 121, available, among other places, at the *Washington Post* website, http://www.washingtonpost.com/wp-srv/national/longterm/unabomber/manifesto.text.htm, accessed January 29, 2016.

It's worth noting that ambivalence regarding the technological project has become increasingly evident among the historians who have dedicated themselves to study of that project. Two essays addressing the subject, one by Brooke Hindle, the other by Carroll W. Pursell Jr., are collected in *The History of American Technology: Exhilaration or Discontent?*, ed. David Hounshell (Wilmington, Del.: Hagley Museum and Library, 1984). Both agree that, beginning in the 1960s, an early emphasis within the academy on the benefits of technology gave way to more critical perspectives. Especially contentious was the assumption that technology serves as a vehicle of social progress. A concluding commentary by Stuart Leslie sums up the prevailing view. "So it seems that our field has abruptly changed tacks in the last couple of decades," he writes, "no longer simply boasting of the fulfillment of America's technological dreams but instead paying more careful attention to how those dreams were fulfilled, what they cost, and who paid the price" (28).

7. Thomas Jefferson, *Notes from the State of Virginia* (1785), query 19, available at American Studies at the University of Virginia website, http://xroads.virginia.edu/~hyper/jefferson /ch19.html, accessed January 29, 2016. The definitive source on the debate during the early days of the republic on whether to pursue manufactures is John F. Kasson, *Civilizing the Machine: Technology and Republican Values in America, 1776–1900* (New York: Hill and Wang, 1976). My paragraph here is largely based on Kasson's opening argument. See also Hugo A. Meier, "Technology and Democracy, 1800–1860," *Mississippi Valley Historical Review* 43, no. 4 (March 1957): 618–40.

8. Kasson, *Civilizing the Machine*, 22–23; and Hugo A. Meier, "Thomas Jefferson and a Democratic Technology," in *Technology in America: A History of Individuals and Ideas*, 2nd ed., ed. Carroll W. Pursell Jr. (Cambridge, Mass.: MIT Press, 1996), 17–33.

9. Jennifer Clark, "The American Image of Technology from the Revolution to 1840," *American Quarterly* 39, no. 3 (Autumn 1987): 439.

10. Leo Marx, *The Machine in the Garden: Technology and the Pastoral Ideal in America* (New York: Oxford University Press, 1964). Simply put, *The Machine in the Garden* is one of the great, irreplaceable works on the history of technology in America and also one of the most quoted.

11. Clark, "American Image of Technology," 439.

12. Marx, *Machine in the Garden*, 136–41.

13. Perry Miller, "The Responsibility of Mind in a Civilization of Machines," an essay in a book of Miller's essays collected under the same title, ed. John Crowell and Stanford J. Searl Jr. (Amherst: University of Massachusetts Press, 1979), 198.

14. See, for example, Robert H. Wiebe, *The Search for Order, 1877–1920* (Boston: Hill and Wang, 1966); Paul A. Carter, *The Spiritual Crisis of the Gilded Age* (DeKalb: Northern Illinois University Press, 1971), and T. J. Jackson Lears, *No Place of Grace: Antimodernism and the Transformation of American Culture, 1880–1920* (New York: Pantheon Books, 1981). Miller's comment quoted by Stuart Leslie in *The History of American Technology*, 28.

15. James quotes from Leo Marx, *Machine in the Garden*, 352–53, and John F. Kasson, *Civilizing the Machine*, 188. Stein's famous remark regarding Oakland is from her autobiography, *Everybody's Autobiography* (New York: Vintage Books, 1973), 289. It refers specifically to the fact that the home in which she had spent much of her youth had literally disappeared. "The house where she grew up was on a sprawling 10-acre plot surrounded by orchards and farms," says Oakland writer Matt Werner. "By 1935, it had been replaced by dozens of houses. . . . Oakland had urbanized and changed from the pastoral place she remembered." See "Gertrude Stein Puts the 'There' back in Oakland," *Google Books*

Search, February 3, 2012, http://booksearch.blogspot.com/2012/02/gertrude-stein-puts
-there-back-in.html, accessed September 14, 2015.

16. Perry Miller, *The Life of the Mind in America: From the Revolution to the Civil War* (New
York: Harcourt, Brace, 1965), 300.

17. Ralph Waldo Emerson, "The Spirit of the Times," in *The Selected Lectures of Ralph Waldo
Emerson*, ed. Ronald A. Bosco and Joel Myerson (Athens: University of Georgia Press,
2005), 124. Bosco and Myerson note that Emerson delivered this lecture in various forms
between 1848 and 1856 and later incorporated significant parts of it into other lectures
as late as 1871.

18. Henry Nash Smith, *Virgin Land: The American West as Symbol and Myth* (1950; repr.,
Cambridge, Mass.: Harvard University Press, 1978), 53–54.

19. Ibid., chap. 6.

20. Thomas Parke Hughes, "The Electrification of America: The System Builders," *Technology and Culture* 20, no. 1 (January 1979): 124–39. See also Hughes's shorter summary
article, "Thomas Alva Edison and the Rise of Electricity," in Pursell, *Technology in America*, 117–28, and Hughes's *American Genesis: A History of the American Genius for Invention* (New York: Penguin, 1989), 86–95. See also Alan Trachtenberg, *The Incorporation
of America: Culture and Society in The Gilded Age* (1982; repr., New York: Hill and Wang,
2007), 65–69.

21. John William Ward, "Charles A. Lindbergh: His Flight and the American Ideal," in Pursell, *Technology in America*, 211–26. This is a brilliant article that describes in detail the
sorts of ambivalence I describe in this chapter.

22. Lindbergh's comparison of the *Spirit of St. Louis* to "a living creature" is from his second
autobiographical account of his flight to Paris, *The Spirit of St. Louis* (New York: Charles
Scribner's Sons, 1953), 486.

23. Ward, "Charles A. Lindbergh," 222, 226.

It's interesting to note that similar sets of contradictory projections defined the
American space program four decades later. According to historian David Mindell,
throughout the Mercury, Gemini, and Apollo missions the astronauts fought a long and
mostly losing campaign to retain control over the spacecraft they ostensibly flew. The
general attitude behind the scenes was that they served mainly as "backup systems,"
on hand to take over in case of malfunction. This was not a story NASA's publicists believed would win taxpayer support. They kept the spotlight on the test pilots turned star
voyagers, reassuring the public that, as Mindell put it, "the classical American hero—
skilled, courageous, self-reliant—had a role to play in a world increasingly dominated
by impersonal technology systems." See Mindell, *Digital Apollo: Human and Machine in
Spaceflight* (Cambridge, Mass.: MIT Press, 2011), 5.

The press responded enthusiastically to that theme. John Glenn's orbital spaceflight
in 1962, for example, provoked an outburst of national elation, not as frenzied as that accorded Lindbergh, but substantial nonetheless. Still, the *New York Times* went out of its
way in an editorial to define the event not as a triumph of science and technology but as
a triumph of human beings over science and technology. "We need not be ruled by machines," the editorial insisted. "We remain the master of the inanimate world and nothing we can make or imagine we can make will take dominion over us." (See "Let Man
Take Over," *New York Times*, February 26, 1962.) The desire of the astronauts to remain
pilots with some degree of control over their spacecraft was, of course, the theme of

Tom Wolfe's masterpiece of new journalism, *The Right Stuff*. Mindell's more recent book is a more straightforward, more detailed account. It is a brilliant piece of scholarship.

24. James L. Fink, "Henry Ford and the Triumph of the Automobile," in Pursell, *Technology in America*, 188. Roderick Nash, *The Nervous Generation: American Thought, 1917–1930* (Chicago: Rand McNally, 1970), chap. 5. Also see Hughes, *American Genesis*, 307, 309.

25. Nash, *Nervous Generation*, 157. When reading that Ford "wrote" an editorial, it should be kept in mind that most of the writing that appeared under his byline was actually composed by his chief public relations man, who can be assumed to have made it his business to accurately reflect the thinking of his boss.

26. Robert Lacey, *Ford, the Men and the Machine* (Boston: Little, Brown, 1986), 228.

27. My description of Greenfield Village is based largely on Lacey, *Ford*, chap. 14. See also Ford's official website on the Greenfield Village theme park, http://bit.ly/18FbRtw, accessed April 2, 2011.

28. Lacey, *Ford*, 242, 245.

29. Ibid., 244.

Chapter 3. At the Intersection of Technology Street and Liberal Arts Avenue

1. Walter Isaacson, *Steve Jobs* (New York: Simon and Schuster, 2011), 526–27.

2. Ibid., 567–68.

3. Ibid., 172–78.

4. Ibid., 179. It's interesting to note that Walter Isaacson is, as of this writing, president and CEO of the Aspen Institute, an institution that grew directly out of the war between science and the humanities. Two of the key figures in the founding of the institute were Robert Hutchins and Mortimer Adler of the University of Chicago, who insisted that science and technology, in their narrow-minded focus on results, were ignoring and eclipsing the foundational wisdom of Western culture, wisdom accrued over the centuries from its classic works of literature and philosophy. This belief was the rationale behind their creation of the famous "Great Books" program, and it was the guiding philosophy of the Aspen Institute at its formation. See James Sloan Allen, *The Romance of Commerce and Culture: Capitalism, Modernism, and the Chicago-Aspen Crusade for Cultural Reform* (Chicago: University of Chicago Press, 1983).

5. Richard Holmes, *Coleridge: Darker Reflections, 1804–1834* (London: Flamingo/Harper Collins, 1998), 492. See also David Newsome, *Two Classes of Men: Platonism and English Romantic Thought* (New York: St. Martin's Press, 1972), chap. 1.

6. William James, *Pragmatism: A New Name for Some Old Ways of Thinking* (New York: Longmans, Green, 1907), 12, 20. All quotations from James are from the opening lecture in this series of lectures.

7. The definitions here are taken from pages 66–67 of *Zen and the Art of Motorcycle Maintenance* (New York: Bantam, 1975). All page numbers refer to Bantam paperback's 2nd printing.

8. James's description of his tough-minded and tender-minded colleagues and Pirsig's description of the classic/romantic split echo polarities that have been repeatedly described through the history of philosophy. Arthur Lovejoy, for example, described the two major strains in Western philosophy since Plato as "this-worldliness" and "other-worldliness," while Hannah Arendt traced the equally ancient dialectic between the *vita*

activa and the *vita contemplativa.* Kant wrote of the differences between practical and theoretical reason, an opposition that subsequently appeared in the works of Thomas Carlyle (Mechanical versus Dynamical natures), Matthew Arnold (Hebraism versus Hellenism), Martin Heidegger (calculative versus meditative reason), and Paul Tillich (technical versus ontological reason).

Nietzsche used imagery from Greek mythology to distinguish between the Apollonian and the Dionysian drives or impulses while Alfred North Whitehead explored the differences between the Reason of Ulysses and the Reason of Plato. Ulysses in Whitehead's formulation represents the inventive pragmatist who comes up with a variety of stratagems to solve real-world problems, compared to Plato's more theoretical intelligence, aimed at finding more general, more transcendent truths. These archetypes in turn bear more than a passing resemblance to Isaiah Berlin's hedgehog and fox (based on a comment by the Greek poet Archilochus), the fox defined as the wily trickster whose energies are directed outward toward the accomplishment of a multitude of goals, the hedgehog as a more inner-directed creature guided by "a single, universal, organizing principle" (*The Hedgehog and the Fox: An Essay on Tolstoy's View of History* [1953; Chicago: Elephant Paperbacks, 1993], 3).

In the 1950s Norman Mailer compiled a long, loose list of opposing characteristics comparing the hip and the square, a polarity that Pirsig also used as another way to describe the classic/romantic split. Mailer introduced his list by saying he'd long wrestled with the idea of writing an entire book on the hip and the square, which he said would have been "a most ambitious *Das Kapital* of the psychic economy." Too ambitious, he decided, considering the energy it would have taken away from his fiction. See *Advertisements for Myself* (New York: G. P. Putnam's Sons, 1959), 424–25.

9. Pirsig, *Zen*, 67.

10. Isaacson, *Steve Jobs*, 397.

11. Three books document from different perspectives the tensions between creativity and control in computer programming. In *The War of Desire and Technology at the Close of the Mechanical Age* (Cambridge, Mass.: MIT Press, 1996), Allucquére Rosanne Stone notes the work of anthropologist Barbara Joans, who described programmers as typically fitting into one of two categories: "Creative Outlaw Visionaries" or "Law and Order Practitioners." Says Stone, "One group fools around with technology and designs fantastic stuff; the other group gets things done and keeps the wheels turning. They talk to each other, if they talk to each other, across a vast conceptual gulf" (14).

In *The Computer Boys Take Over: Computers, Programmers, and the Politics of Technical Expertise* (Cambridge, Mass.: MIT Press, 2010), Nathan Ensmenger traces the power struggles between programmers and managers that unfolded inside American corporations as computers became increasingly important to their operations from the 1950s onward. The independent, nonconformist, often antisocial personalities of many of the early programmers brought them into conflict with their superiors, who hoped to supervise and control the programmers without having the slightest understanding of what they did. These tensions also played out in engineering circles as associations, would-be certifiers, and individuals attempted to define and enforce "professionalism" in their burgeoning field.

In *Engineers for Change: Competing Visions of Technology in 1960s America* (Cambridge, Mass.: MIT Press, 2012), Matthew Wisnioski describes a series of attempts during the countercultural era to bridge the classic/romantic split. Not a few engineers (though still a minority of the profession) were swept up in the currents of the time and hoped to infuse their work with more artistic, more humanist values. The results, Wisnioski reports, were disappointing. I discuss the counterculture's relationship to technology in detail in chapter 5.

12. Isaacson, *Steve Jobs*, 48–49.
13. Lynn White Jr., "The Historical Roots of Our Ecologic Crisis," *Science* 155, no. 3767 (March 10, 1967): 1203–7. Max Weber, *The Protestant Ethic and the Spirit of Capitalism* (1905).
14. David Newsome, *Two Classes of Men*, 4.
15. A sign over the entrance to Plato's academy supposedly read, "Let no one enter who has not studied geometry." For a representative example of his thoughts on geometry, see *The Republic*, book 7, 526c–527c. There Socrates dismisses geometry's more practical applications, as in war and business, calling them inferior to its value as a vehicle for perceiving "the Form of the good." Properly employed, he says, geometry directs the mind toward "knowledge of that which always is, not of that which at some particular place and time is becoming and perishing." It has the power, he adds, to "draw the souls toward truth and complete the philosophic understanding, making us raise upwards what we now wrongly direct downwards." Translated by A. D. Lindsay, Everyman's Library ed. (Rutland, Vt.: J. M. Dent/Charles E. Tuttle, 1992).
16. Carl Mitcham, *Thinking through Technology: The Path between Engineering and Philosophy* (Chicago: University of Chicago Press, 1994), 277–82.

    Leo Strauss echoed this view. The classic Greek philosophers, he wrote, "were for almost all practical purposes what are now called conservatives. In contradistinction to many present day conservatives, however, they knew that one cannot be distrustful of political or social change without being distrustful of technological change. . . . They demanded the strict moral-political supervision of inventions; the good and wise city will determine which inventions are to be made use of and which are to be suppressed." From *Thoughts on Machiavelli*, quoted in Mulford Q. Sibley, "Utopian Thought and Technology," *American Journal of Political Science* 17, no. 2 (May 1973): 258.
17. Hannah Arendt, *The Human Condition* (Chicago: University of Chicago Press, 1958), 15. See Arendt's discussion of Greek attitudes on technology on pp. 12–16 and 153–59. Arendt's points are echoed by Robert C. Scharff and Val Dusek in their introduction to part 1, *Philosophy of Technology: The Technological Condition, an Anthology* (Malden, Mass.: Blackwell, 2003), 3–5.
18. Bacon, *Novum Organum*, ed. Joseph Devey (Forgotten Books; repr. of New York: P. F. Collier and Son, 1901), 49, 58.

Chapter 4. Antipathies of the Most Pungent Character

1. William James, *Pragmatism: A New Name for Some Old Ways of Thinking* (New York: Longmans, Green, 1907), 12–13.
2. A compilation of Apple's commercials portraying a clueless Microsoft is available on YouTube at https://www.youtube.com/watch?v=C5zoIa5jDt4. The IBM "1984" commercial is available on YouTube at https://www.youtube.com/watch?v=2zfqw8nhUwA. Both accessed December 26, 2015.
3. Swift's satire on this subject is in the Voyage to Laputa section of *Gulliver's Travels*; see chapter 10 for a description of that section. For an account of the worshipful following inspired by Bacon, see Richard Foster Jones's *Ancients and Moderns: A Study of the Rise*

of the Scientific Movement in Seventeenth-Century England (Berkeley: University of California Press, 1965). Mumford's comment is taken from an introduction he wrote to an edition of Samuel Butler's *Erewhon and Erewhon Revisited* (New York: Modern Library, 1927), xxi.

4. The quote "some sort of bizarre sexual satisfaction" is from Robert Graysmith, *Unabomber: A Desire to Kill* (New York: Berkley, 1998), 144. On Kaczynski's waste disposal and supply precautions, see Chris Waits and Dave Shors, *Unabomber: The Secret Life of Ted Kaczynski* (Helena: Helena Independent Record and Montana Magazine, 1999), 236–38.

5. For the quote on Kaczynski's debates with his brother David, see Alston Chase, *Harvard and the Unabomber: The Education of an American Terrorist* (New York: Norton, 2003), 332.

6. I'm grateful to Stefan Collini, whose introduction to the Canto reprint edition of C. P. Snow's "Two Cultures" lecture 13 brought the debate between Huxley and Arnold to my attention. See Snow, *The Two Cultures* (Cambridge: Cambridge University Press, 1998), xiii–xv.

7. Mason Science College was subsequently incorporated into the University of Birmingham. Information on Josiah Mason and the college can be found at the university's website: http://bit.ly/18FcCCT, accessed March 31, 2011.

   Regarding the birth of mass-market consumerism in England, the historian Neil McKendrick has written, "There was a consumer boom in England in the eighteenth century. In the third quarter of the century that boom reached revolutionary proportions. Men, and in particular women, bought as never before. Even their children enjoyed access to a greater number of goods than ever before. In fact, the later eighteenth century saw such a convulsion of getting and spending, such an eruption of new prosperity, and such an explosion of new production and marketing techniques, that a greater proportion of the population than in any previous society in human history was able to enjoy the pleasures of buying consumer goods." From McKendrick, John Brewer, and J. H. Plumb, *The Birth of a Consumer Society: The Commercialization of Eighteenth-Century England* (Bloomington: Indiana University Press, 1982), 9.

8. Huxley's lecture, "Science and Culture," is available online at http://bit.ly/18FcLGj, accessed March 31, 2011, and is drawn from his *Collected Essays*, 9 vols. (London: Methuen, 1893–1902).

9. Ibid.

10. Arnold's lecture is available online at http://bit.ly/1a2WXuX, accessed March 31, 2011.

    Huxley delivered the Rede Lecture the year following Arnold's. The title of his address was "The Origin of the Existing Forms of Animal Life: Construction or Evolution?"

11. Emerson quoted in John F. Kasson, *Civilizing the Machine: Technology and Republican Values in America, 1776–1900* (New York: Hill and Wang, 1976), 118.

12. Whitehead, *Science and the Modern World* (1925; repr., New York: Free Press, 1967), 77, 84, 94.

13. It's equally odd that neither Huxley nor Arnold mentioned in their speeches the Quarrel of the Ancients and the Moderns, an exchange that unfolded among French and English intellectuals in the late seventeenth and early eighteenth centuries. Here, too, the bone of contention in the later argument was virtually identical to that of its predecessor, i.e., whether the discoveries of modern science had superseded the wisdom of the classical philosophers. Jonathan Swift's *Battle of the Books*, written in 1697 and published in 1704,

was a satire on that debate. For a detailed account of this fascinating historical turning point, see Richard Foster Jones's *Ancients and Moderns*.

14. Snow, *The Two Cultures*, 4, 11–12, 15–17.

15. Ibid., 10–11, 22.

16. Ibid., 53–100. These pages contain Snow's follow-up essay, "The Two Cultures: A Second Look," originally published 1963.

17. Stephen Jay Gould, *The Hedgehog, the Fox, and the Magister's Pox: Mending the Gap Between Science and the Humanities* (Cambridge, Mass.: Harvard University Press, 2011), 89–90. As its title suggests, Gould directly addressed in this book (his last) the classic/romantic split. He spent a fair amount of space responding to Edward O. Wilson's *Consilience: The Unity of Knowledge*, an earlier treatise also aimed at reconciling the two poles. Gould objected to many of Wilson's ideas but emphatically agreed with his conviction that the greatest intellectual challenge facing the species "has always been and always will be finding a way to combine or connect our different ways of knowing" (2–3).

18. F. R. Leavis, *Two Cultures? The Significance of C. P. Snow* (New York: Cambridge University Press, 2013). Again, I am indebted to Stefan Collini's superb introduction to the Canto edition of Snow's lecture, which lays out much of the background covered here.

19. Ibid., 56, 60.

20. Ibid., 54–55.

21. Ibid., 70–71.

22. Ibid., 72.

23. Ibid., 73.

Chapter 5. A Momentary Interruption

1. Histories that trace the counterculture's role in the computer revolution include John Markoff's *What the Dormouse Said: How the Sixties Counterculture Shaped the Personal Computer Industry* (New York: Penguin, 2005) and Fred Turner's *From Counterculture to Cyberculture: Stewart Brand, the Whole Earth Network and the Rise of Digital Utopianism* (Chicago: University of Chicago Press, 2006).

   One of the best depictions of the troubled relationship between technology and the counterculture can be found in a book that takes as its subject the attention devoted during that period to healing the breach between them. As mentioned in an earlier note (see chap. 3, note 11) in *Engineers for Change: Competing Visions of Technology in 1960s America* (Cambridge, Mass.: MIT Press, 2012), Matthew Wisnioski documents how engineers who were sympathetic to counterculture views tried to encourage the adoption of more humanistic, more socially conscious values by their profession. Numerous companies, universities, and professional organizations initially supported these efforts, in large part because they were anxious to defend themselves against the widespread perception of America's youth that technology was a force of evil. "What is it like to be an engineer," asked one prominent engineer, "at the moment that the profession has achieved unprecedented successes and simultaneously is being accused of having brought our civilization to the brink of ruin?" (Wisnioski, *Engineers for Change*, 5, quoting Samuel C. Forman's *The Existential Pleasures of Engineering*.)

   Although significant discussion, debate, and energy were devoted to addressing those concerns, Wisnioski shows that relatively few engineers actively participated. "Across every member society and in the public relations of the corporations that employed their members, engineers were urged to assume responsibility for the social-technical world," Wisnioski writes. In the end, however, the "great majority" of these efforts amounted to "more heat than light." Adding courses on the social impacts of technology to engineering school curricula also failed to significantly alter the status quo, Wisnioski says. As a 1973 report by MIT's Center for Policy Alternatives put it, "A decade of protest and turmoil has left engineering campuses and students only slightly changed." (See pp. 5, 89–90, 185.)

2. Carl Mitcham, *Thinking through Technology: The Path between Engineering and Philosophy* (Chicago: University of Chicago Press, 1994), 290–91.

3. "The Port Huron Statement of the Students for a Democratic Society," available at: http://coursesa.matrix.msu.edu/~hst306/documents/huron.html, accessed March 13, 2016.

4. Books reflecting the counterculture's mistrust of technology include Charles A. Reich's *The Greening of America* (New York: Random House, 1970) and Theodore Roszak's *Where the Wasteland Ends: Politics and Transcendence in Post-Industrial Society* (Garden City, N.Y.: Anchor Books, 1973).

   Because it was originally published in the *New Yorker*, Reich's *The Greening of America* is probably the best-known critique of technology to appear in the sixties. Perhaps the most penetrating, however, is John McDermott's 1969 essay in the *New York Review of Books*, "Technology: The Opiate of the Intellectuals." The piece was inspired by McDermott's outrage over an earlier essay, "The Social Impact of Technological Change," by Emmanuel G. Mesthene, director of Harvard University's Program on Technology and Society. Both essays are reprinted in *Philosophy of Technology: The Technological Condition, an Anthology*, ed. Robert C. Scharff and Val Dusek (Malden, Mass.: Blackwell, 2003), 617–37 (Mesthene) and 638–51 (McDermott).

   Another sign of the times was "Technology and Pessimism," a three-day seminar convened by leading scholars of technology who were concerned about the era's general negativity toward machines. See John Noble Wilford, "Scholars Confront the Decline of Technology's Image," *New York Times*, November 6, 1979.

5. Theodore Roszak, *From Satori to Silicon Valley: San Francisco and the American Counterculture* (San Francisco: Don't Call It Frisco Press, 1986), 9. Note that I've changed Roszak's spelling of "tipi" to the more conventional "teepee."

6. See note 8 in chapter 3 citing Norman Mailer's list of opposing characteristics comparing the hip and the square.

7. Robert M. Pirsig, *Zen and the Art of Motorcycle Maintenance* (New York: Bantam, 1975), 16.

8. Jobs occasionally used the word "grok," a word coined by Heinlein in *Stranger in a Strange Land*. Here is one of Heinlein's definitions, according to Wikipedia: "Grok means to understand so thoroughly that the observer becomes a part of the observed—to merge, blend, intermarry, lose identity in group experience."

9. Andrew Kirk, *Counterculture Green: The Whole Earth Catalog and American Environmentalists* (Lawrence: University Press of Kansas, 2007), 54, 57.

10. My account of the space-colonies debate is taken from the two issues of the *CoEvolution Quarterly* (*CEQ*) in which it was prominently featured, Fall 1975 and Spring 1976. I also referred to *Space Colonies* (New York: Penguin, 1977), a collection of material from that debate edited by Stewart Brand. The quotation from Stewart Brand in this paragraph

is from *CEQ*, Fall 1975, 6. The quotation from Drexler is from *Space Colonies*, 104. For a history of the L5 Society, see http://www.nss.org/settlement/L5news/L5history.htm, accessed March 28, 2016.

11. *CEQ*, Fall 1975, 9–11.

12. Ibid., 12, 17.

13. Ibid., 5. It's interesting to note that in the book version of this passage, the wording was changed to read that "the most dogmatic Space Colony proponents" expected the project to solve the energy crisis, the food crisis, the arms race, and the population problem. The book also added that the space colonies might solve an additional problem: "Climate Shift." See *Space Colonies*, 6–7.

14. *CEQ*, Fall 1975, 4–5.

15. Ibid., 29.

16. Ibid., 5–6. I don't recall the name of Frederick Jackson Turner ever surfacing in the space-colonies debate. Nonetheless, the ambitions of O'Neill and his supporters underscore the thesis of Turner's paper "The Significance of the Frontier in American History," delivered to the American Historical Association in 1893.

17. *CEQ*, Spring 1976, 5.

18. Ibid.

19. Ibid., 6, 10.

20. Ibid., 15, 29.

21. Ibid., 8.

22. Ibid.

23. Ibid., 9.

24. The historian David E. Nye has identified the "technological foundation story" as one of the central mythologies of the American experience. These stories follow a characteristic pattern, Nye says: A group or individual enters an undeveloped region with one or more new technologies; part of the region is transformed through use of the new technologies; the new settlement prospers, and more settlers arrive; land values increase, and some settlers become wealthy; largely shaped by the new technologies, the original landscape disappears and is replaced by a "second creation" that completes or improves the original, natural creation; the process begins again as some members of the community depart for another undeveloped region. See *America as Second Creation: Technology and Narratives of New Beginnings* (Cambridge, Mass.: MIT Press, 2003), 9, 13.

25. *CEQ*, Spring 1976, 53.

26. Brand, *Space Colonies*, 85.

27. Ibid., 82.

28. Ibid., 85.

29. On Musk, see Ross Andersen, "Exodus," *Aeon*, September 30, 2014, http://aeon.co/magazine/technology/the-elon-musk-interview-on-mars/, accessed September 22, 2015. On Bezos, see promotional information from his company, Blue Origin, at https://www.blueorigin.com/news and at https://www.blueorigin.com/#youtubekbT29lA322g, accessed November 27, 2015.

30. Jonathan Miles, "The Billionaire King of Techtopia," *Details*, September 2011, http://bit.ly/oHiLJr, accessed August 22, 2013. Thiel subsequently began to have second thoughts about the plausibility of his aquatic utopia, in part because he and his partners realized that the scale of disruption they would encounter on the high seas would be of a different order of magnitude than what they were used to in Silicon Valley. As the Seasteading

Institute's revised mission statement puts it, "The high cost of open ocean engineering serves as a large barrier to entry and hinders entrepreneurship in international waters. This has led us to look for cost-reducing solutions within the territorial waters of a host nation." See Kyle Denuccio, "Silicon Valley Is Letting Go of Its Techie Island Fantasies," *Wired*, May 16, 2015, http://www.wired.com/2015/05/silicon-valley-letting-go-techie -island-fantasies/, accessed September 22, 2015.

Chapter 6. What Is Technology?

1. David Foster Wallace began his famous commencement speech at Kenyon College with a version of the fish story and also includes it in his novel *Infinite Jest*.
2. Thomas Parke Hughes, *American Genesis: A History of the American Genius for Invention* (New York: Penguin, 1989), 5, emphasis in original.
3. Ibid., 6. Mitcham's definition is quoted by Albert Borgmann in *Technology and the Character of Contemporary Life: A Philosophical Inquiry* (Chicago: University of Chicago Press, 1984), 13. An abbreviated version of that definition appears on page 1 of Mitcham's *Thinking through Technology: The Path between Engineering and Philosophy* (Chicago: University of Chicago Press, 1994). Edward Tenner, *Our Own Devices: The Past and Future of Body Technology* (New York: Knopf, 2003), ix.
4. Joseph Pitt, *Thinking about Technology* (New York: Seven Bridges Press, 2000), xi, 11.
5. For a more complete overview of Ellul's ideas and personal history, see my profile of him in the *Boston Globe*, "Jacques Ellul: Technology Doomsayer before His Time," July 8, 2012, http://bo.st/VuBAiy, accessed July 31, 2013.
6. Ellul, *The Technological Society* (New York: Vintage, 1964), 4.
7. See *Broad and Narrow Interpretations of Philosophy*, ed. Paul T. Durbin, *Philosophy of Technology*, vol. 7 (London: Kluwer Academic, 1990). Joseph Pitt's essay in that volume, where he defines his terms "internalist" and "externalist," is "In Search of a New Prometheus," 3–15. See also Carl Mitcham's distinction between humanities philosophy of technology versus engineering philosophy of technology, *Thinking through Technology*, chaps. 1, 2, and 6.

   John M. Staudenmaier, SJ, uses the word "contexturalist" to refer to those who integrate consideration of the design of a technology with some aspect of its context. As the longtime editor of the journal *Technology and Culture*, Staudenmaier has calculated that the vast majority of articles published by that journal reflect a contexturalist perspective. See his essay "Rationality versus Contingency in the History of Technology," in *Does Technology Drive History? The Dilemma of Technological Determinism*, ed. Merritt Roe Smith and Leo Marx (Cambridge, Mass.: MIT Press, 1998), 267–68, and also his book, *Technology's Storytellers: Reweaving the Human Fabric* (Cambridge, Mass.: MIT Press, 1985).
8. Thomas Misa, "Retrieving Sociotechnical Change," in Smith and Marx, *Does Technology Drive History?*, 141. See also Stephan Kline, "What Is Technology?," in *Philosophy of Technology: The Technological Condition, an Anthology*, ed. Robert Scharff and Val Dusek (Malden, Mass.: Blackwell, 2003), 210–12.

   In regard to what constitutes "hardware," the question may arise: What is the difference between a tool and a machine? According to Witold Rybczynski, "it is generally agreed that tools are instruments that are manually operated, while machines transmit some sort of external force." From *Taming the Tiger: The Struggle to Control Technology* (New York: Viking, 1983), 61.

9. Quoted by David Mindell, *Digital Apollo: Human and Machine in Spaceflight* (Cambridge, Mass.: MIT Press, 2011), 37.

10. Lewis Mumford, *Technics and Human Development* (San Diego: Harcourt Brace Jovanovich, 1966), 196, 189, and chap. 9. Also see Mumford, "History: Neglected Clue to Technological Change," *Technology and Culture* 2, no. 3 (Summer 1961): 232.

11. Regarding "Industrial Revolution": Lewis Mumford and others object to the use of that label because they believe it suggests, falsely, that substantial industrial development did not occur prior to the eighteenth century. For evidence to the contrary, see Jean Gimpel's *The Medieval Machine: The Industrial Revolution of the Middle Ages* (New York: Penguin, 1977). I use the term because I subscribe to the conventional view that the scale and pervasiveness of technological development beginning in the eighteenth century represented a break with previous periods.

12. Joseph Pitt, "The Autonomy of Technology," in *Philosophy and Technology*, vol. 3, *Technology and Responsibility*, ed. Paul T. Durbin (Dordrecht, Netherlands: D. Reidel, 1987), 99–113.

13. Jacques Ellul, *The Ethics of Freedom* (Grand Rapids, Mich.: William B. Eerdmans, 1976), 40–41. Langdon Winner provides an excellent defense of Ellul on this score. See *Autonomous Technology: Technics-out-of-Control as a Theme in Political Thought* (Cambridge, Mass.: MIT Press, 1977), 227–28.

14. From Lewis Mumford, *Technics and Civilization* (San Diego: Harcourt Brace, 1934). Used as an epigraph by David F. Noble in his book *America by Design: Science, Technology, and the Rise of Corporate Capitalism* (New York: Oxford University Press, 1979). For a similar statement by Mumford written nearly thirty years later, see "History: Neglected Clue to Technological Change," 230.

15. For an exposition of this point, see Langdon Winner's "Do Artifacts Have Politics?," *Daedalus* 109, no. 1 (Winter 1980): 121–36.

16. Hegel discusses this principle at length in *The Encyclopedia of Philosophical Sciences*. See part 1, "The Doctrine of Being." The idea that at some point a change in quantity becomes a change in quality is formally known as "incrementalism." Informally, the phrases "the tipping point" and "the straw that broke the camel's back" describe essentially the same phenomenon.

17. The relationship between science and technology has been the subject of voluminous discussion in academic circles. Useful overviews include Otto Mayr, "The Science-Technology Relationship as a Historiographic Problem," *Technology and Culture* 17, no. 4 (October 1976): 663–73; George Wise, "Science and Technology," *Osiris* 1, Historical Writing on American Science (1985), 229–46; Edwin Layton Jr., "Through the Looking Glass, or News from Lake Mirror Image," *Technology and Culture* 28, no. 3 (July 1987), and "Mirror-Image Twins: The Communities of Science and Technology in 19th Century America," *Technology and Culture* 12, no. 4 (October 1971): 562–80. For a news article that does a good job of describing the science-technology issue, see William J. Broad, "Does Genius or Technology Rule Science?," *New York Times*, August 7, 1984, http://nyti.ms /13moV3M, accessed July 31, 2013.

18. World's Fair motto and Henderson quoted by Thomas J. Misa, *Leonardo to the Internet* (Baltimore: Johns Hopkins University Press, 2011), 300–301. Price quoted by George Wise

in his *Osiris* article cited in n. 17 of this chapter, p. 229. Price's prose is, in general, abstruse, but for one of his representative articles on this subject, see "Is Technology Historically Independent of Science? A Study in Statistical Historiography," *Technology and Culture* 6, no. 4 (Autumn 1965): 553–68.

19. Cyril Stanley Smith, "Art, Technology, and Science: Notes on Their Historical Interaction," *Technology and Culture* 11, no. 4 (October 1970): 498.

20. Layton, "Through the Looking Glass," 603; Mayr, "Science-Technology Relationship," 667–68. For a concise description of the interactive model of the relationship between science and technology, see Barry Barnes, "The Science-Technology Relationship: A Model and a Query," *Social Studies of Science* 12, no. 1 (February 1982): 166–72.

     The interrelationships of technology, art, and science are regularly explored in the well-known TED conferences (the name is an acronym for technology, entertainment, and design). For an article on the ongoing interaction of these fields, see Amy Wallace, "Science to Art, and Vice Versa," *New York Times*, July 9, 2011, http://nyti.ms/nVbOTD, accessed July 31, 2013.

21. Ellul, *Technological Society*, 7–10.

22. "Engineering Biology: A Talk with Drew Endy," *Edge.com*, 2008, http://bit.ly/18E16Zg, accessed July 6, 2011.

23. Ellul, *Technological Society*, 311–12.

24. Noble, *America by Design*, xxv. Alfred D. Chandler, *The Visible Hand: The Managerial Revolution in American Business* (1977; repr., Cambridge, Mass.: Harvard University Press, 1999), xxv. Note that in *The Coming of Post-Industrial Society* (1973; New York: Basic Books, 1999) Daniel Bell sidestepped this question (whether cleverly or evasively, I'm not sure) with his description of a socially decisive "techno-economic" realm (xix).

Chapter 7. The Nature of Technology

1. The word "nature" has many meanings, which can lead to confusion. The cultural historian Arthur O. Lovejoy outlined a taxonomy of those meanings in connection with his studies of the Neo-Classical and Romantic movements of the seventeenth and eighteenth centuries. See "'Nature' as Aesthetic Norm," in *Essays in the History of Ideas* (Baltimore: Johns Hopkins University Press, 1948), 69–77.

     My description of the "nature" of technology corresponds to the definition of nature Lovejoy described as "the system of necessary and self-evident truths concerning the properties and relations of essences." Lovejoy also cited uses of "nature" that refer to "the average type, or statistical 'mode' of a kind." This definition, he added, implied to some degree that "nature" represents the realization in empirical reality of "the essence or Platonic Idea of a kind." These, too, have some bearing on my use of the word.

2. I need to acknowledge that the interpretations here are my own. I also should acknowledge that my interpretations may qualify me as an "advocacy historian" as defined by the historian of technology Brooke Hindle. "Advocacy historians of whatever stamp," he said, "seek to apply the historian's integrity to their politically oriented reading of history." This is partially true in my case. I don't seek to apply the historian's integrity to my own reading of history but rather to draw on the historian's research and insights to inform my own perceptions. Admittedly, there's a thin line between the two. My approach is more in tune with the practices of journalism, a craft I've practiced for many years. Although I do appropriate in these pages the works of many historians, I attempt to do so

as fairly as possible. Hindle's comments are contained in his essay "The Exhilaration of American Technology: A New Look," in *The History of American Technology: Exhilaration or Discontent?*, ed. David A. Hounshell (Wilmington, Del.: Hagley Museum and Library, 1984), 12.

3. Isaiah Berlin, *The Hedgehog and the Fox: An Essay on Tolstoy's View of History* (1953; Chicago: Elephant Paperbacks, 1933), 4.

4. The historian Donald MacKenzie has called the question of technological determinism "perhaps *the* central question" in the history of technology. In the next chapter I'll explain the close relationship between determinism and autonomy. MacKenzie's comment is from his article, "Marx and the Machine," *Technology and Culture* 25, no. 3 (July 1984): 499.

5. Langdon Winner, *The Whale and the Reactor: A Search for Limits in an Age of High Technology* (Chicago: University of Chicago Press, 1989), 174, emphasis in original.

6. Jacques Ellul, *The Technological Society* (New York: Vintage, 1964), 85.

7. The name McDonald's supplies a popular textbook's descriptive shorthand for the sorts of issues examined here. See *The McDonaldization of Society*, by George Ritzer (Thousand Oaks, Calif.: Pine Forge Press, 2000).

8. Langdon Winner, *Autonomous Technology: Technics-out-of-Control as a Theme in Political Thought* (Cambridge, Mass.: MIT Press, 1977), 210. Francis Bacon, *New Atlantis and The Great Instauration*, ed. Jerry Weinberger (Wheeling, Ill.: Harlan Davidson, 1989), 71.

9. Google, "The New Multi-Screen World: Understanding Cross-Platform Consumer Behavior," http://bit.ly/17EQLaX, accessed March 17, 2013. Sam Costello, "How Many Apps Are There in the App Store?," About.com, http://tinyurl.com/bup5f9u, accessed March, 14, 2016.

10. Ted Kaczynski addressed the incentive toward technological expansion in his manifesto. Paragraph 132 reads: "It is well known that people generally work better and more persistently when striving for a reward than when attempting to avoid a punishment or negative outcome. Scientists and other technicians are motivated mainly by the rewards they get through their work. But those who oppose technological invasions of freedom are working to avoid a negative outcome, consequently there are a few who work persistently and well at this discouraging task. If reformers ever achieved a signal victory that seemed to set up a solid barrier against further erosion of freedom through technological progress, most would tend to relax and turn their attention to more agreeable pursuits. But the scientists would remain busy in their laboratories, and technology as it progresses would find ways, in spite of any barriers, to exert more and more control over individuals and make them always more dependent on the system."

11. Ellul, *Technological Society*, 142, 193.

12. Lewis Mumford, *Technics and Civilization* (San Diego: Harcourt Brace, 1934), 65–77; Carolyn Merchant, *The Death of Nature: Women, Ecology and the Scientific Revolution* (San Francisco: Harper, 1980), 170–72. In *Paradise Lost*, Milton portrays Satan using mining, smelting, forging, and molding to construct the city of Pandemonium. See Carl Mitcham, *Thinking through Technology: The Path between Engineering and Philosophy* (Chicago: University of Chicago Press, 1994), 294.

13. Martin Heidegger, "Memorial Address," in *Discourse on Thinking* (New York: Harper and Row, 1966), 46. My comment that Heidegger was notorious for "other things" besides his idiosyncratic language is a reference to his Nazism, an appalling fact of his life that is at odds with the brilliance of his perspectives on technology. I make no attempt to explain

that disjunction and certainly make no attempt to excuse it. It should be noted, however, that he considered Jews especially prone to "calculative" thinking. See Peter E. Gordon, "Heidegger in Black," *New York Review of Books*, October 9, 2014.

14. Martin Heidegger, "The Question Concerning Technology," in *Basic Writings*, ed. David Farrell Krell (New York: Harper and Row, 1977), 292–93, 296–99.

15. Marshall McLuhan, *Understanding Media: The Extensions of Man* (1964; repr., Cambridge, Mass.: MIT Press, 1998), 73, 85. Also see pp. 11–12.

16. The definitive biography of Taylor is Robert Kanigel's *The One Best Way: Frederick Winslow Taylor and the Enigma of Efficiency* (New York: Penguin, 1997). Also see Samuel Haber, *Efficiency and Uplift: Scientific Management in the Progressive Era, 1890–1920* (Chicago: University of Chicago Press, 1964).

   It's interesting to note that while Taylor considered the stopwatch technique the "keystone" of scientific management, it was soon superseded by a motion-picture technique pioneered by his one-time associate and eventual rival, Frank Gilbreth. "Micromotion study" allowed Gilbreth to break down worker movements to a thousandth of a second. See Brian Price, "Frank and Lillian Gilbreth and the Motion Study Controversy, 1907–1930," in *A Mental Revolution: Scientific Management since Taylor*, ed. Daniel Nelson (Athens: Ohio University Press, 1992), 58–76.

17. Kanigel, *One Best Way*, 169. Despite his recognition of the efficiencies the divisions of labor could produce, Adam Smith also recognized that their imposition necessarily required the sort of numb acquiescence that Taylor preferred. As he put it in *The Wealth of Nations*, a man who spends his life performing the same, simple operation over and over again "generally becomes as stupid and ignorant as it is possible for a human creature to become." See Smith, *An Inquiry into the Nature and Causes of the Wealth of Nations* (Edinburgh: Thomas Nelson, 1743), 321.

18. Drucker quotation in Douglas Brinkley, *Wheels for the World: Henry Ford, His Company, and a Century of Progress* (New York: Viking, 2003), 140. On efficiency as the essence of Fordism, see Brinkley, *Wheels for the World*, 141. On Ford's lack of knowledge of Taylor's writings, see the review by Alfred D. Chandler Jr. of Allen Nevins's *Ford: The Times, the Man, the Company*, *Business History Review* 28, no. 4 (December 1954): 387–89.

19. Clayton Christensen, "We Are Living the Capitalist's Dilemma," CNN, January 21, 2013, http://bit.ly/WH8bRi, accessed March 17, 2013. For an analysis of the impact of efficiency measures on labor, see Josh Bivens and Lawrence Mishel, "Understanding the Historic Divergence between Productivity and a Typical Worker's Pay: Why It Matters and Why It's Real," Economic Policy Institute, September 2, 2015, http://tinyurl.com/nhfyyj5, accessed September 30, 2015.

   It should be noted that what constitutes efficiency differs in the eye of the beholder. For example, to turn supervision of a national meat inspection program over to meat producers may be efficient as far as the producers are concerned, but not necessarily as far as meat consumers are concerned. See Ted Genoways, "Making a Pig's Ear of Food Safety," *New York Times*, December 12, 2014, http://tinyurl.com/zlh2mo, accessed December 6, 2015.

20. Mailer, *Of a Fire on the Moon*, included in the collection *The Time of Our Time* (New York: Random House, 1998), 718.

21. The historian Raul Hilberg has described the Nazis' prosecution of the Holocaust as an example of convergence, combining the technique of the concentration camp with the technology of the gas chamber: "As separate establishments, both the concentration

camp and the gas chamber had been in existence for some time. The great innovation was [put into] effect when the two devices were fused." See *The Destruction of the European Jews*, 3rd ed., vol. 3 (1961; repr., New Haven, Conn.: Yale University Press, 2003), 922.

22. Winner, *Autonomous Technology*, 238; Ray Kurzweil, *The Singularity Is Near: When Humans Transcend Biology* (New York: Penguin, 2005), 134–35.

For the past decade or so many scientists have been advocating the conscious pursuit of convergence as the research and development strategy of the future. See, for example, "The Third Revolution: The Convergence of the Life Sciences, Physical Sciences and Engineering," paper presented in 2011 by a group of research scientists at MIT. It is available online at http://bit.ly/144tnXu. An earlier report, "Converging Technologies for Improving Human Performance: Nanotechnology, Biotechnology, Information Technology and Cognitive Science," was sponsored by the National Science Foundation and published in 2003. It is available online at http://bit.ly/12kPmIq. Both accessed August 1, 2013.

23. On the application of bronze casting techniques to cannons, see Witold Rybczynski, *Taming the Tiger: The Struggle to Control Technology* (New York: Viking, 1983), 9–10. On the application of steam technology to railroads, see Joseph Weizenbaum, *Computer Power and Human Reason: From Judgment to Calculation* (San Francisco: W. H. Freeman, 1976), 32.

24. Examples of products derived from the space program are taken from David Baker, *Scientific American: Inventions from Outer Space; Everyday Uses for NASA Technology* (New York: Random House, 2000) and from Marlowe Hood and Laurent Banguet, "Heavenly Gadgets: Spinoffs from Space Programmes," AFP/Phys.Org, April 10, 2011, http://bit.ly/1bjKxCY, accessed May 11, 2011. On the use of video-game processors in the military's Roadrunner computer, see John Markoff, "Military Supercomputer Sets Record," *New York Times*, June 9, 2008, http://nyti.ms/11wGeOc, accessed July 31, 2013.

Note that John Staudenmaier makes a distinction between technological diffusion and the transfer of technology. He defines diffusion as a process taking place within the culture of origin and transfer as a technology exported to another "recipient" culture or cultures. See *Technology's Storytellers: Reweaving the Human Fabric* (Cambridge, Mass.: MIT Press, 1985), 123.

25. On the influence of slaughterhouses on Ford's assembly lines, see Sigfried Giedion, *Mechanization Takes Command: A Contribution to Anonymous History* (1948; repr., New York: W. W. Norton, 1969), 77–78. On McDonald's applying assembly-line techniques to hamburgers, see Eric Schlosser, *Fast Food Nation: The Dark Side of the American Meal* (New York: HarperCollins, 2002), 20, 68–69.

David S. Landes has pointed out that a "direct chain of innovation" led from the earliest factories of the Industrial Revolution to modern assembly lines simply because the complexities of factories made the efficient organization of labor and materials a pressing and consistent concern. See *The Unbound Prometheus: Technological Change and Industrial Development in Western Europe from 1750 to the Present* (Cambridge: Cambridge University Press, 1969), 2.

Regarding the fact that the disassembly processes of Chicago slaughterhouses helped inspire Henry Ford's concept of the assembly line, the historian Richard Holmes has noted that Mary Shelley described a similar reversal nearly a century earlier in *Frankenstein*. In that case the disassembly of corpses in the dissection theaters of the period was turned in Shelley's imagination into a laboratory in which dissected body parts were assembled into a living creature. See Holmes, *The Age of Wonder* (New York: Vintage, 2008), 327.

26. On technological disequilibrium, see Nathan Rosenberg, "Technological Change in the Machine Tool Industry, 1840–1910," *Journal of Economic History* 23, no. 4 (December 1963): 440. On the chain of innovations in the textile industry, see John F. Kasson, *Civilizing the Machine: Technology and Republican Values in America, 1776–1900* (New York: Hill and Wang, 1976), 21–22. On the development of the telegraph as a signaling system for railroads, see Wolfgang Schivelbusch, *The Railway Journey: Trains and Travel in the 19th Century* (Berkeley: University of California Press, 1986), 36–39.

   Note that while Rosenberg is credited with originating the concept of technological disequilibrium, Jacques Ellul had anticipated it as early as 1954. "Each new machine," he wrote, "disturbs the equilibrium of production; the restoration of equilibrium entails the creation of one or more additional machines in other areas of operation." *Technological Society*, 112.

   For a brief explanation of the concept of "social lag" as it relates to technological convergence and expansion, see George H. Daniels, "The Big Questions in the History of American Technology," *Technology and Culture* 11, no. 1 (January 1970): 2–3. The classic work on cultural lag is William F. Ogburn's *Social Change with Respect to Culture and Original Nature* (New York: B. W. Huebsch, 1922). For a critical review of the concept, see Joseph Schneider, "Cultural Lag: What Is It?," *American Sociological Review* 10, no. 6 (December 1945): 786–91.

   In regard to slavery, historians have argued that the pervasiveness of slavery in premodern societies was a disincentive to the development of technology in those societies. As Lewis Mumford put it, "As long as human bone and sinew were more plentiful than metal, no other form of machine was called for." See "Tools and the Man," *Technology and Culture* 1, no. 4 (Autumn 1960): 323.

27. For a brilliant study of the development of the American system, see Merritt Roe Smith, *Harpers Ferry Armory and the New Technology: The Challenge of Change* (Ithaca, N.Y.: Cornell University Press, 1977). Smith and other historians have demolished the myth that Eli Whitney was the sole progenitor of the American system. See Robert S. Woodbury, "The Legend of Eli Whitney and Interchangeable Parts," *Technology and Culture* 1, no. 3 (Summer 1960): 235–53.

28. Rosenberg, "Technological Change in the Machine Tool Industry," 425, 426, 430. See also Rosenberg's introduction in a collection of documents he edited, *The American System of Manufactures* (Edinburgh: Edinburgh University Press, 1969); and David Hounshell, *From the American System to Mass Production* (Baltimore: Johns Hopkins University Press, 1985).

   Note that while Rosenberg's article in the *Journal of Economic History* is justly considered an important contribution to American technological as well as economic history, it is not, contrary to his suggestion, the first place the term "technological convergence" was used. In fact, Jacques Ellul had discussed the phenomenon, using the same terminology, nine years earlier in *The Technological Society* (see p. 391). Rosenberg can be forgiven for not being aware of Ellul's usage, however; *The Technological Society*, originally published in France in 1954, wasn't published in the United States until 1964.

29. Bacon quoted by Alan Kors, "Bacon's *New Organon: The Call for a New Science*," audio lecture in the series "The Great Minds of the Western Intellectual Tradition," part 2 (Chan-

tilly, Va.: Teaching Company, 1992). The quotation from Tocqueville is from the chapter in *Democracy In America* titled "Why the Americans Are More Addicted to Practical Than to Theoretical Science," quoted by Meier, "Technology and Democracy, 1800–1860," *Mississippi Valley Historical Review* 43, no. 4 (March 1957): 624.

30. Edwin T. Layton Jr., *The Revolt of the Engineers: Social Responsibility and the American Engineering Profession* (Baltimore: Johns Hopkins University Press, 1986), 3. For discussion of the importance of chemistry and electricity in encouraging the synthesis of science and technology, see David F. Noble, *America by Design: Science, Technology, and the Rise of Corporate Capitalism* (New York: Oxford University Press, 1979), 5. For the importance of precision measurement in provoking that growing synthesis, see p. 71.

31. The quotation from Taylor is from Layton, *Revolt of the Engineers*, 139–40.

32. Quoted by Thomas Parke Hughes, *American Genesis: A History of the American Genius for Invention* (New York: Penguin, 1989), 73.

    David A. Hounshell has documented the sometimes prickly relationship between Edison and the scientific community of his era. Edison at times relished his identification with and acceptance into that community and at other times chafed at its constraints. See "Edison and the Pure Science Ideal in 19th Century America," *Science*, n.s., 207, no. 4431 (February 8, 1980): 612–17.

    It's worth noting that Edison represents fulfillment of a corporate approach to scientific discovery and technological application that was envisioned in the seventeenth century by Francis Bacon. As Lewis Mumford put it, Bacon "felt that science in the future would rest increasingly on a collective organization, not just on the work of individuals of ability, operating under their own power. . . . He saw that curiosity, to be fully effective, must enlist, not solitary and occasional minds, but a corps of well-organized workers, each exercising a specialized function and operating in a restricted area." See "Science as Technology," *Proceedings of the American Philosophical Society* 105, no. 5 (October 13, 1961): 506, 508.

33. On the growing alliance of science and industry, see Hughes, *American Genesis*, and Noble's *America by Design*, especially chaps. 1 and 2. Both books are brilliant works of historical synthesis. Willis Whitney quoted by Daniel Boorstin, *The Republic of Technology: Reflections on Our Future Community* (New York: Harper and Row, 1978), 27.

34. My account of the Cadillac interchangeability challenge is based on Maurice D. Hendry, *Cadillac, Standard of the World: The Complete History* (New York: Automobile Quarterly/E. P. Dutton, 1977), 49–51.

    Note that the "challenge" from the Royal Automobile Club was actually provoked by Cadillac's British representative at the time, Frederick Stanley Bennett, who suggested the interchangeability test as a means to publicly demonstrate Cadillac's mechanical superiority. Also note that I am unsure whether the challenge was actually issued in late 1907 or early 1908. According to Hendry, Bennett talked to members of the British trade press about the test in October 1907 and the test itself began on February 29, 1908.

    Background information on Henry Leland is from Rosenberg, "Technological Change," and from Jeff McVey, "A History of the Leland & Faulconer Manufacturing Co.," http://vintagemachinery.org/mfgindex/detail.aspx?id=1462, accessed March 31, 2011. For information on Leland's experience in the manufacture of firearms, see Yanek Mieczkowski, "The Man Who Brought Us Cadillac and Lincoln," History News Network, George Mason University, http://hnn.us/articles/1646.html, accessed March 31, 2011.

Note as well that Henry Leland was involved with one of the early incarnations of the Ford Motor Company, the Henry Ford Company. He was brought in by Ford's investors to take over from Ford as chief engineer in hopes that Leland would force Ford to stop experimenting with different prototypes and get a product on the market. The two men were too strong-willed to coexist, and Ford soon departed. The Henry Ford Company soon became the Cadillac Motor Company, named after Detroit's founder, Antoine de la Mothe Cadillac. See Douglas Brinkley, *Wheels for the World*, 41–43.

35. The last sentence in this paragraph is an adaptation of a comment by the scientist and historian of science J. D. Bernal, who said: "The cardinal tendency of progress is the replacement of an indifferent chance environment by a deliberately created one." Quoted by David F. Noble, *The Religion of Technology* (New York: Alfred A. Knopf, 1997), 175.

36. Gregory Stock, *Redesigning Humans: Our Inevitable Genetic Future* (Boston: Houghton Mifflin, 2002), 2.

37. On the 1965 blackout, see James Burke, *Connections* (Boston: Little, Brown, 1978), 1–2.

38. Edison quoted by Hughes, *American Genesis*, 73. On the subject of technological control in general, see James R. Beniger, *The Control Revolution: Technological and Economic Origins of the Information Society* (Cambridge, Mass.: Harvard University Press, 1986), and Miriam R. Levin, ed., *Cultures of Control* (New York: Routledge, 2004). The concept of "tightly coupled systems" is from Charles Perrow, *Normal Accidents: Living with High-Risk Technology* (New York: Basic Books, 1984).

39. See Alfred D. Chandler, *Visible Hand: The Managerial Revolution in American Business* (1977; repr., Cambridge, Mass.: Harvard University Press, 1999), chap. 3. Chandler documents how a fatal collision of Western Railroad passenger trains in 1841 prompted the company to adopt not only a rigorous system of timetables but also "the first modern, carefully defined, internal organization structure used by an American business enterprise" (97).

See also Ian R. Bartky, "The Adoption of Standard Time," *Technology and Culture* 30, no. 1 (January 1989): 25–56. Bartky argues that the direct impetus for the imposition of standardized time zones in the United States was provided not by the railroads but by astronomers, meteorologists, and geodesists who needed consistent standards for purposes of accurate measurement. Nonetheless, he also provides ample evidence that, although many railroad executives (not all) were satisfied with maintaining their companies' independent time schedules, many observers, including the U.S. Naval Observatory, had been pressing for a common "railway time" long before a standardized national time system was adopted. Bartky also shows that it was railroad superintendents and managers who actually implemented such a system, agreeing to do so in part to forestall the imposition by government of a time scheme less to their liking. Standard railway time went into effect in November 1883; the U.S. Congress adopted the system in March 1884.

40. The eclipse of the independent farmer began sooner than is often suspected. Rather than a late twentieth-century phenomenon (and inspiration for the series of musical benefits known as "Farm Aid"), the disappearance of the family farm was well under way by the latter third of the nineteenth century, hastened by a combination of corporate farming, the opening of global markets, the high price of machinery, speculation on commodity exchanges, rapacious fees for storing and transporting harvested crops, and usurious credit. See Steve Fraser, *The Age of Acquiescence: The Life and Death of American Resistance to Organized Wealth and Power* (Boston: Little, Brown, 2015), chap. 4.

On gigantism in the tech industry, see Farhad Manjoo, "Tech's 'Frightful 5' Will Dominate Digital Life for Foreseeable Future," *New York Times*, January 20, 2016, http://tinyurl.com/jm9gdak; and Don Clark and Robert McMillan, "Facebook, Amazon, and Other Tech Giants Tighten Grip on Internet Economy," *Wall Street Journal*, November 5, 2015, www.wsj.com/articles/giants-tighten-grip-on-internet-economy-144677132. Both accessed March 24, 2016.

41. Ellul had much to say about the techniques of corporate human relations, which serve, he said, as "a kind of lubricating oil" to adapt the individual to the technological milieu. See *Technological Society*, 337–56.

42. Ibid., 135–36.

43. Jodi Kantnor and David Streitfeld, "Inside Amazon: Wrestling Big Ideas in a Bruising Workplace," *New York Times*, August 15, 2015, http://tinyurl.com/j8efhjh, accessed March 3, 2016. See also Simon Head, "Worse Than Wal-Mart: Amazon's Sick Brutality and Secret History of Ruthlessly Intimidating Workers," *Salon*, February 23, 2014, http://tinyurl.com/n5g3xdg, accessed March 3, 2016.

For my essay on the broader implications of Amazon's work ethic, see "Amazon.com: Apotheosis of Technique," http://thequestionconcerningtechnology.blogspot.com/2015/08/amazoncom-apotheosis-of-technique.html, accessed September 6, 2015.

44. Ellul, *Technological Society*, 142–43.

45. Kantor and Streitfeld, "Inside Amazon."

46. John Higham, "Hanging Together: Divergent Unities in American History," *Journal of American History* 61, no. 1 (June 1974): 18–19.

Chapter 8. Who's in Charge Here?

1. It must be stated at the outset of the notes for this chapter that Langdon Winner's *Autonomous Technology: Technics-out-of-Control as a Theme in Political Thought* (Cambridge, Mass.: MIT Press, 1977) is the indispensable examination of this subject.

2. Boorstin, "Bicentennial Essay: Tomorrow: The Republic of Technology," *Time*, January 17, 1977. Boorstin later expanded this essay into a book, *The Republic of Technology: Reflections on Our Future Community* (New York: Harper and Row, 1978).

3. Quote is from *The New Industrial State*, cited by Winner, *Autonomous Technology*, 14.

4. Werner Heisenberg, *Physics and Philosophy: The Revolution in Modern Science* (Amherst, N.Y.: Prometheus Books, 1999), 189. Heisenberg's view that technological expansion had become a "biological" process has subsequently been echoed and expanded by others, among them the well-known technology writer Kevin Kelly. See *What Technology Wants* (New York: Viking, 2010).

5. Karl Marx, *The Poverty of Philosophy* (Chicago: Charles H. Keer, 1910), 119. Note that whether Marx can accurately be called a technological determinist is a question of spirited debate among scholars. See Donald MacKenzie, "Marx and the Machine," *Technology and Culture* 25, no. 3 (July 1984): 473–502, and William H. Shaw, "'The Handmill Gives You the Feudal Lord': Marx's Technological Determinism," *History and Theory* 18, no. 2 (May 1979): 155–76.

6. An excellent guide to academic discussion of these issues is *Does Technology Drive History? The Dilemma of Technological Determinism*, ed. Merritt Roe Smith and Leo Marx (Cambridge, Mass.: MIT Press, 1998). See especially the thoughtful essays by the two editors and by John Staudenmaier. The title of the book alludes to the title of a key pa-

per on the subject by Robert L. Heilbroner, included in the collection, "Do Machines Make History?" See also Allan Dafoe's more recent review of the determinism issue, "On Technological Determinism: A Typology, Scope Conditions, and a Mechanism," *Science, Technology, & Human Values* 40, no. 6 (2015): 1047–76. Dafoe proposes that determinism should be "reclaimed from its use as a critic's term and straw position" within academic debate (1049). For a debate on social constructivism and technology, see Langdon Winner's essay "Social Constructivism: Opening the Black Box and Finding it Empty"; Mark Elam's response to that essay, "Anti Anticonstructivism or Laying the Fears of a Langdon Winner to Rest"; and Winner's response to Elam, all reprinted in *Philosophy of Technology: The Technological Condition, an Anthology*, ed. Robert Scharff and Val Dusek (Malden, Mass.: Blackwell, 2003).

For other helpful discussions of the subject of technological determinism, see Langdon Winner's *Autonomous Technology*, chap. 2, and his article "Do Artifacts Have Politics?," *Daedalus* 109, no. 1 (Winter 1980): 121–36. Also helpful is Reinhard Rürup, "Historians and Modern Technology: Reflections on the Development and Current Problems of the History of Technology," *Technology and Culture* 15, no. 2 (April 1974): 161–93; and George H. Daniels, "The Big Questions in the History of American Technology," *Technology and Culture* 11, no. 1 (January 1970): 1–21. See also John Staudenmaier, *Technology's Storytellers: Reweaving the Human Fabric* (Cambridge, Mass.: MIT Press, 1985), 134–48.

7. Steven Levy, "Technomania," *Newsweek*, February 27, 1995. Levy continued the theme sixteen years later with his 2011 book, *In the Plex: How Google Thinks, Works and Shapes Our Lives* (New York: Simon and Schuster, 2011).

8. Abraham Lincoln, in his annual address to Congress in 1862, said: "That portion of the earth's surface which is owned and inhabited by the people of the United States is well adapted to be the home of one national family, and it is not well adapted for two or more. Its vast extent and its variety of climate and productions are of advantage in this age for one people, whatever they may have been in former ages. Steam, telegraphs and intelligence have brought these to be an advantageous combination for one united people." Quoted by Roger Burlingame, "Technology: Neglected Clue to Historical Change," *Technology and Culture* 2, no. 3 (Summer 1961): 228.

9. The assumption that technology and science are what separate us from the savages was a central theme of the Enlightenment, from Francis Bacon on. See Robert Scharff and Val Dusek, *Philosophy of Technology: The Technological Condition, an Anthology* (Malden, Mass.: Blackwell, 2003), 27; and Rosalind Williams, "Nature Out of Control: Cultural Origins and Environmental Implications of Large Technological Systems," in *Cultures of Control*, ed. Miriam R. Levin (New York: Routledge, 2004), 45.

10. See pp. 47–60 of Jacques Ellul, *The Technological Society* (New York: Vintage, 1964). A plethora of texts address the question of America's technological precociousness, from Alexis de Tocqueville's *Democracy in America* (see chap. 9) to Daniel Boorstin's *The Americans: The National Experience* (New York: Vintage, 1965). See, for example, Boorstin's remark that Americans excelled in technical discovery simply because they were willing to try anything. "Ignorance and 'backwardness' had kept Americans out of the old grooves. Important innovations were made because Americans did not know any better" (21). For the argument that republican values prompted a drive for economic independence from England, see John F. Kasson, *Civilizing the Machine: Technology and Republican Values in America, 1776–1900* (New York: Hill and Wang, 1976). Regarding the varying degrees of social plasticity in different regions of the United States, see Rich-

ard D. Brown, *Modernization: The Transformation of American Life, 1600–1865* (Prospect Heights, Ill.: Waveland Press, 1976), esp. 114, 120. See also Merritt Roe Smith, *Harpers Ferry Armory and the New Technology: The Challenge of Change* (Ithaca, N.Y.: Cornell University Press, 1977). For an analysis of how social plasticity accounted for Europe's leadership in the development of industrialism, see the introduction in David S. Landes, *The Unbound Prometheus: Technological Change and Industrial Development in Western Europe from 1750 to the Present* (Cambridge: Cambridge University Press, 1969), 12–33. For discussion of how social and political disruption set the stage for Germany's prosecution of the Holocaust (which was, among other things, a technological phenomenon), see Zygmunt Bauman, *Modernity and the Holocaust* (1989; repr., Ithaca, N.Y.: Cornell University Press, 2000), 111.

11. Regarding the differences in technical intention between Islam and the West, see Benjamin R. Barber, *Jihad vs. McWorld: Terrorism's Challenge to Democracy* (New York: Ballantine, 1995). Regarding the Native Americans, see Jerry Mander's *In the Absence of the Sacred: The Failure of Technology and the Survival of the Indian Nations* (San Francisco: Sierra Club Books, 1991).

   Note that a combination of technical intention and social plasticity may help explain why the city of Birmingham, England (where Josiah Mason founded his Science College) was a seedbed of the Industrial Revolution. According to historian Jenny Uglow, Birmingham was known for its independence, its religious tolerance, and its ambition. "A city of makers and traders, Birmingham almost seemed itself to be 'in the making,' always looking forward. Many people claimed that part of the reason for its growth was its freedom from rules. It had no charter to shackle it, and no ancient craft guilds to block enterprise with strict apprenticeship and trading rules." See *The Lunar Men: Five Friends Whose Curiosity Changed the World* (New York: Farrar, Straus and Giroux, 2002), 19.

12. Winner, *Autonomous Technology*, 75. Carl Mitcham discusses the inherent power of having a tool versus not having a tool in *Thinking through Technology: The Path between Engineering and Philosophy* (Chicago: University of Chicago Press, 1994), 183. See also Hannah Arendt's discussion of instrumentality in *The Human Condition*, 151–57. Note, too, Saul Bellow's comment regarding technology as a form of projection: "All artifacts originate in thought. They are thoughts practically extended into matter." From the collection of Bellow's essays, *There Is Simply Too Much to Think About*, ed. Benjamin Taylor (New York: Penguin, 2015), 261.

13. Winner, *Autonomous Technology*, 100–101, emphasis in the original.

14. Ibid., 88–90. Henry Adams recognized the constraints of the technological imperative more than a century ago. The expansion of the American railroad system in the nineteenth century was absorbing the attention and energy of the nation, he wrote, "for it required all the new machinery to be created—capital, banks, mines, shops, powerhouses, technical knowledge, mechanical population, together with a steady remodeling of social and political habits, ideas, and institutions to fit the new scale and suit the new conditions. The generation between 1865 and 1895 was already mortgaged to the railways, and no one knew it better than the generation itself." *The Education of Henry Adams: An Autobiography* (1907; repr., Boston: Houghton Mifflin, 2000), 240.

15. Hughes, "Technological Momentum," in *Does Technology Drive History? The Dilemma of Technological Determinism*, ed. Merritt Roe Smith and Leo Marx (Cambridge, Mass.: MIT Press, 1998), 101–13. This is an updated essay on the ideas originally articulated in

Hughes, "Technological Momentum: Hydrogenation in Germany 1900–1933," *Past and Present* 44, no. 1 (August 1969): 106–32.

16. On Western Union and Bell, see David A. Hounshell, "On the Disadvantages of Being an Expert," *Technology and Culture* 16, no. 2 (April 1975): 157. There are many accounts of Steve Jobs's fateful visit to Xerox Parc. See, for example, Michael Moritz, *Return to the Little Kingdom: Steve Jobs, the Creation of Apple, and How It Changed the World* (New York: Overlook Press, 1999), 306–9.

17. John Markoff, *What the Dormouse Said: How the Sixties Counterculture Shaped the Personal Computer Industry* (New York: Penguin, 2005), 135. On Google's struggle to remain nimble, see Steven Levy, "Larry Page Wants to Return Google to Its Startup Roots," *Wired*, April 2001, http://bit.ly/hRHdTu. Also see Clair Cain Miller, "Google Grows, and Works to Retain Nimble Minds," *New York Times*, November 28, 2010, http://nyti.ms/gTR301; Miller and Miguel Helft, "Google Shake-Up Is Effort to Revive Start-Up Spark," *New York Times*, January 20, 2011, http://nyti.ms/dTLDFx; and Jessica Guynn, "Larry Page Is fostering Google's Start-Up Spirit," *Los Angeles Times*, April 1, 2011, http://lat.ms/gGcYlT. All articles accessed August 1, 2013.

It's well known that preconceived ideas can hold back innovative development in fields of thought and endeavor beyond technology, including politics, ethics, and religion. This was a fundamental conviction underlying the philosophy of John Dewey, who wrote that habits of belief and character create a "lag" in social progress much as developmental lag slows the progress of technological development. The four "idols" identified by Francis Bacon addressed similar tendencies as blocks to scientific reason.

18. Ellul, *Technological Society*, 85–94.

19. Edwin Layton, "The Interaction of Technology and Society," *Technology and Culture* 11, no. 1 (January 1970): 29. Note that Layton's view fits the position that Albert Borgmann describes as "pluralist." See *Technology and the Character of Contemporary Life: A Philosophical Inquiry* (Chicago: University of Chicago Press, 1984), 9–10.

20. For Mitcham's comments on whether society or technology is the "primary member" in the relationship, see *Thinking through Technology*, 275.

21. Regarding the values and meaning built into technological systems, see Mitcham, *Thinking through Technology*, 252, and Langdon Winner's "Do Artifacts Have Politics?"

Chapter 9. Quality

1. Pirsig's notion of Quality is related to the philosopher Albert Borgmann's more rigorous concept of "focal things and practices." Borgmann argues that by consciously focusing on "tangible and bodily things" we can overcome the vacuity encouraged by technological convenience and ubiquity. His discussion uses the experiences of running and dining ("the table") as examples of how focal things and practices can restore conscious awareness to our interactions with the technological environment. "The mind becomes relatively disembodied when the body is severed from the depth of the world, i.e., when the world is split into commodious surfaces and inaccessible machineries. . . . If we are to challenge *the rule of technology*, we can do so only through *the practice of engagement*." See *Technology and the Character of Contemporary Life: A Philosophical Inquiry* (Chicago: University of Chicago Press, 1984), 196–210, emphases in the original.

2. Pirsig, *Zen*, 27.

3. John Lachs, *Responsibility and the Individual in Modern Society* (Brighton, England: Harvester Press, 1981), 32.

4. Matthew B. Crawford, *Shop Class as Soulcraft: An Inquiry into the Value of Work* (New York: Penguin Press, 2009), 2, 7, 61.

5. Walter Isaacson, *Steve Jobs* (New York: Simon and Schuster, 2011), 71, 75, 137–38.

6. Mihaly Csikszentmihalyi and Eugene Rochberg-Halton, *The Meaning of Things: Domestic Symbols and the Self* (Cambridge: Cambridge University Press, 1981), 4–8.

   There is an affinity between Csikszentmihalyi and Rochberg-Halton's ideas and the emphasis on feedback in Norbert Wiener's theory of cybernetics. Their ideas also resonate with the Chinese practice of *feng shui* and with Miranda Lambert's country music hit "The House That Built Me."

7. Csikszentmihalyi and Rochberg-Halton, *Meaning of Things*, 117.

   Herbert Marcuse made the same observation with a more pejorative slant: "The people recognize themselves in their commodities; they find their soul in their automobile, hi-fi set, split-level home, kitchen equipment." *One Dimensional Man: Studies in the Ideology of Advanced Industrial Society* (Boston: Beacon Press, 1964), 9.

   In 1982, the French philosopher Michel Foucault described "technologies of the self," which he defined as technologies that permit individuals "to effect by their own means or with the help of others a certain number of operations on their own bodies and souls, thoughts, conduct, and way of being, so as to transform themselves in order to attain a certain state of happiness, purity, wisdom, perfection, or immortality." *Technologies of the Self: A Seminar with Michel Foucault*, ed. Luther H. Martin, Huck Gutman, and Patrick H. Hutton (Amherst: University of Massachusetts Press, 1988), 18.

8. Csikszentmihalyi and Rochberg-Halton, *Meaning of Things*, 9–10. Pirsig, *Zen*, 298–312.

9. Csikszentmihalyi and Rochberg-Halton, *Meaning of Things*, 28.

10. Davis Baird, "Thing Knowledge—Function and Truth," *Techné: Research in Philosophy and Technology* 6, no. 2 (Winter 2002): 96–105. Ford quote from Greg Grandin, *Fordlandia: The Rise and Fall of Henry Ford's Forgotten Jungle City* (New York: Henry Holt, 2009), 256.

11. Csikszentmihalyi and Rochberg-Halton, *Meaning of Things*, 58–59. The authors add that because of the scarcity of personal possessions in the homes of even the wealthy during the Middle Ages, furniture became a symbol of power. Hence the throne of royalty and the designation of a group leader as the "Chairman." On the "Consumer Revolution" that developed in parallel with the Industrial Revolution, see John McKendrick, John Brewer, and J. H. Plumb, *The Birth of a Consumer Society: The Commercialization of Eighteenth-Century England* (Bloomington: Indiana University Press, 1982).

12. De Zengotita, "The Numbing of the American Mind: Culture as Anesthetic," *Harpers*, April 2002, 33–40. See also De Zengotita's *Mediated: How the Media Shapes Your World and the Way You Live in It* (New York: Bloomsbury, 2005).

13. Marshall McLuhan, *Understanding Media: The Extensions of Man* (1964; repr., Cambridge, Mass.: MIT Press, 1998), 47.

14. Ruskin quoted by Wolfgang Schivelbusch, *The Railway Journey: Trains and Travel in the 19th Century* (Berkeley: University of California Press, 1986), 57–58.

15. From *Life Is a Miracle*, quoted by Glen Mazis, *Humans, Animals, Machines: Blurring Boundaries* (Albany: SUNY Press, 2008), 164.

16. Emerson quoted by Leo Marx, "The Machine in the Garden," *New England Quarterly* 29,

no. 1 (March 1956): 35–36. Frisch quoted by Daniel Boorstin, *The Republic of Technology: Reflections on Our Future Community* (New York: Harper and Row, 1978), 59. The quote comes from Frisch's novel *Homo Faber: A Report*, trans. Michael Bullock (San Diego: Harvest, 1987), German original published in 1957.

    In the *New England Quarterly* article, Marx makes the point that the dislocations described by a wide variety of writers during Emerson's period were "invariably" associated with "science and industrial technology."

17. Lewis Mumford, "Tools and the Man," *Technology and Culture* 1, no. 4 (Autumn 1960): 326. Georg Simmel's comment regarding "the blasé attitude" of city dwellers is from his essay "The Metropolis and Mental Life," in *On Individuality and Social Forms* (Chicago: University of Chicago Press, 1971), 329, emphasis in the original. His comment regarding the intensification of nervous stimulation is quoted by Schivelbusch, *Railway Journey*, 57.

18. Wordsworth: "The Prelude; VII: Residence in London" (ll. 722–28). Note that Wordsworth worked on "The Prelude" for some fifty years and it remained unpublished until three months after his death in 1850. There are slight variations in the text between published editions.

19. Matthew Arnold, "The Scholar-Gipsy," ll. 141–46.

20. Ken Auletta, *Googled: The End of the World as We Know It* (New York: Penguin, 2009), 229.

21. Jenks quoted by James Livingston, "The Social Analysis of Economic History and Theory: Conjectures on Late Nineteenth-Century American Development," *American Historical Review* 92, no. 1 (February 1987): 93. Thanks to Steve Fraser for this reference.

22. Henry Adams, *The Education of Henry Adams: An Autobiography* (1907; repr., Boston: Houghton Mifflin, 2000), 4–5.

23. Émile Durkheim, *Suicide: A Study in Sociology*, trans. John A. Spaulding and George Simpson (New York: Routledge, 1970), 367–69.

    The statistics Durkheim cites were indeed shocking. Between 1826 and 1890, he said, suicide rates in Prussia had risen by 411 percent; in France during almost the same period they rose by 385 percent. Over shorter time spans he found rises in suicides of 212 percent in Belgium and 109 percent in Italy. Durkheim commented that suicide was "most widespread everywhere in the most cultivated regions," to which he added, "What the rising flood of voluntary deaths denotes is not the increasing brilliancy of our civilization but a state of crisis and perturbation not to be prolonged with impunity" (334, 369).

24. Schlesinger quoted by Matthew Wisnioski, *Engineers for Change: Competing Visions of Technology in 1960s America* (Cambridge, Mass.: MIT Press, 2012), 41.

25. On the cult of antiquarianism, see Lewis Mumford, *Technics and Civilization* (San Diego: Harcourt Brace, 1934), 312. On Beard and neurasthenia, see Alan Trachtenberg, *The Incorporation of America: Culture and Society in The Gilded Age* (1982; repr., New York: Hill and Wang, 2007), 47. The description of the symptoms of neurasthenia are from T. J. Jackson Lears, *No Place of Grace: Antimodernism and the Transformation of American Culture, 1880–1920* (New York: Pantheon Books, 1981), 50.

26. On Spencer and *Harper's*, see Lears, *No Place of Grace*, 50–52.

27. For a summary of statistics on contemporary increases of depression and suicide, and on workplace dissatisfaction, see Bruce E. Levine, "Living in America Will Drive You Insane—Literally," *Salon*, July 31, 2013, http://bit.ly/18Noa1n. See also Jesse Singal, "For 80

Years, Young Americans Have Been Getting More Anxious and Depressed, and No One Is Quite Sure Why," *New York*, March 13, 2016, http://tinyurl.com/jmhqenk. On nostalgia, see Kurt Andersen, "So You Say You Want a Devolution?," *Vanity Fair*, January 2012, http://vnty.fr/tqwdhF. All accessed March 25, 2016.

28. On Heidegger, see Carl Mitcham, *Thinking through Technology: The Path between Engineering and Philosophy* (Chicago: University of Chicago Press, 1994), 257. On Hegel, see Nicholas Bunnin and Jiyuan Yu, *The Blackwell Dictionary of Philosophy* (Carlton, Victoria, Australia: Blackwell, 2004),412, and Langdon Winner, *Autonomous Technology: Technics-out-of-Control as a Theme in Political Thought* (Cambridge, Mass.: MIT Press, 1977), 286. Marx quoted by Donald MacKenzie in his article "Marx and the Machine," *Technology and Culture* 25, no. 3 (July 1984), 475.

    For an excellent essay on Heidegger's views on how attention affects perception, see Lawrence Berger, "Being There: Heidegger on Why Our Presence Matters," *New York Times*, March 30, 2015, http://tinyurl.com/mlvoeu8, accessed September 23, 2015.

29. Borgmann, *Technology and the Character of Contemporary Life*, 44–46.

30. Terkel, *Working: People Talk about What They Do All Day and How They Feel about What They Do* (New York: Pantheon Books, 1974), xxxi–xxxii.

31. Ibid., 49–50.

32. Tony Schwartz and Christine Porath, "Why You Hate Work," *New York Times*, May 30, 2014, http://tinyurl.com/oqpttsh; and Schwartz, "You Don't Have to Hate Your Job," June 6, 2014, http://tinyurl.com/hjheyva. Both accessed November 3, 2015.

33. On Taylorism's divorce of labor from thought, see Samuel Haber, *Efficiency and Uplift: Scientific Management in the Progressive Era, 1890–1920* (Chicago: University of Chicago Press, 1964), 24.

34. My essay on the romantic sensibilities that guided Steve Jobs's design and execution of Apple's products can be found at the September 25, 2012, entry, "Steve Jobs, Romantic," on my blog, *The Question Concerning Technology*, http://bit.ly/15y4yOe, accessed September 5, 2013.

35. Tillich, *The Spiritual Situation in the Technological Society*, ed. J. Mark Harris (Macon, Ga.: Mercer University Press, 1988), 41.

Chapter 10. Absorption

1. Steven Levy, *Hackers: Heroes of the Computer Revolution* (1984; repr., New York: Penguin, 1994).

2. Ibid., 33, 42–43.

3. Ibid., 37.

    Regarding the hackers' limited access to mainframe computers, it's interesting to note that a similar situation prevailed at the very outset of the scientific revolution. According to Steven Shapin and Simon Schaffer, authors of *Leviathan and the Air-Pump: Hobbes, Boyle, and the Experimental Life* (Princeton, N.J.: Princeton University Press, 1985), the air pump used by Robert Boyle for his seminal series of scientific experiments was "seventeenth-century Big Science," expensive to build and difficult to build correctly. Only the privileged few had access. A laboratory equipped with one, Shapin and Shaffer write, was "disciplined space, where experimental, discursive, and social practices were collectively controlled by competent members" (38–39).

One can easily speculate how the ability to experiment with these pivotal tools—the air pump and the computer alike—might have influenced the distribution of economic and social benefits derived from that experience.

4. Levy, *Hackers*, 37.

5. Ibid., 37–38.

6. Joseph Weizenbaum, *Computer Power and Human Reason: From Judgment to Calculation* (San Francisco: W. H. Freeman, 1976), 116.

7. Gates, *The Road Ahead* (New York: Viking, 1995), 17.

    Any number of similar descriptions of the hacker mentality could be cited. One example is Katherine Losse's portrait of Thrax, a programmer she knew when she worked at Facebook in its early days (she was employee no. 51). Thrax's schedule, she says, consisted of twenty-hour days on the computer followed by sleep, "from which I imagined him waking only to put his fingers back on the keypad and resume the line of code or AIM chat that he was writing when he passed out." She describes this sort of sleep cycle as "akin to being plugged into an electric socket at all times, minus fresh air, circadian rhythms, or exercise." Losse, *The Boy Kings: A Journey into the Heart of the Social Network* (New York: Free Press, 2012), 67. These characteristics are not dissimilar, of course, from those ascribed to Facebook founder Mark Zuckerberg in the movie *The Social Network*.

8. Levy, *Hackers*, 39.

9. Ibid., 37.

10. Jonathan Swift, *Gulliver's Travels*, Norton Critical Edition (New York: W. W. Norton, 1970), 138. The authoritative analysis of the targets at which the voyage to Laputa was aimed is Marjorie Nicolson and Nora M. Mohler, "The Scientific Background of Swift's Voyage to Laputa," *Annals of Science* 2 (1937): 299–334. According to Nicolson and Mohler, doubts regarding the utility of scientific speculation and the arrogance of those who practiced it were widespread in Swift's day. Other commentators argue that science and scientists were widely respected. Both are probably true. There's no doubt that a massive fad for experimentation swept across Europe, inspired by Francis Bacon. Massive fads commonly provoke satirical responses.

    No doubt some readers will have noticed the similarity between the satirical method of Swift and that of Aristophanes, the playwright of ancient Greece whose comedy "The Clouds" depicts Socrates floating above the Earth in a basket. This illustrates a point that in fairness needs to be acknowledged, i.e., that philosophers and authors can be as lost in thought as scientists and computer programmers.

11. Twain, *Life on the Mississippi* (New York: Penguin, 1984), 94–95. Leo Marx discusses this passage at length in *The Machine in the Garden: Technology and the Pastoral Ideal in America* (New York: Oxford University Press, 1964), 320–24. For a more extensive discussion, see Marx's earlier article, "The Pilot and the Passenger: Landscape Conventions and the Style of Huckleberry Finn," *American Literature* 28, no. 2 (May 1956): 129–46, reprinted in *Mark Twain: A Collection of Critical Essays*, ed. Henry Nash Smith (Englewood Cliffs, N.J.: Prentice-Hall, 1963), 47–63.

12. Twain, *Life on the Mississippi*, 95.

13. Ibid.

14. Marx, *Machine in the Garden*, 320.

15. Charles Lindbergh, *The Spirit of St. Louis* (New York: Charles Scribner's Sons, 1953), 249. I'm grateful to the historian John F. Kasson for alerting me to this passage in Lindbergh's

memoir. See *Civilizing the Machine: Technology and Republican Values in America, 1776–1900* (New York: Hill and Wang, 1976), 116.

16. Lindbergh, *Spirit of St. Louis*, 250.

17. Tom Wolfe, *The Right Stuff* (New York: Bantam, 1980), 269. I'm indebted to Langdon Winner for reminding me of this story; he relates it in the opening paragraph of *The Whale and the Reactor: A Search for Limits in an Age of High Technology* (Chicago: University of Chicago Press, 1989), 3.

18. See Frank White, *The Overview Effect: Space Exploration and Human Evolution*, 2nd ed. (Reston, Va.: American Institute of Aeronautics and Astronautics, 1998).

19. On Ihde, see Carl Mitcham, *Thinking through Technology: The Path between Engineering and Philosophy* (Chicago: University of Chicago Press, 1994), 77.

20. Peter Berger, Brigitte Berger, and Hansfried Kellner, *The Homeless Mind: Modernization and Consciousness* (New York: Random House, 1973), 30–31, 34, 39–40, 125. Regarding the doctrine of separate spheres, see Ruth Schwartz Cowan, *More Work for Mother: The Ironies of Household Technology from the Open Hearth to the Microwave* (New York: Basic Books, 1983), 18–19.

    The need to keep a firm boundary between work life and private life, and the difficulty of doing so, was a prominent theme in Francis Ford Coppola's American epic, *The Godfather*, and in the television series *Breaking Bad*. The late rap star Notorious B.I.G., a former crack dealer, also sounded this theme in his song "Ten Crack Commandments," which advises that a rule frequently "underrated" holds that a dealer must keep his family and his business "completely separated."

21. Reeve Lindbergh, "Childhood Memories of Charles Lindbergh," *Smithsonian*, December 20, 2010. This is an excerpt from Reeve Lindbergh's memoir, *Under a Wing*.

22. Herman Melville, *Moby-Dick, or, The Whale* (New York: Penguin, 1992), 183.

23. Ibid., 182–83. Note that Melville's characterizations of Ahab and Ishmael are masterful representations of the classic/romantic split. For Ahab, in addition to the chapter quoted (chap. 37, "Sunset"), see chap. 28, in which he's described as a man made of solid bronze. Ishmael, by contrast, is a classic dreamer, his head almost literally in the clouds. In chap. 35, "The Mast-Head," he confesses that during his watches in the crow's nest, the "blending cadences of waves and thoughts" lull him into a state of what we would now describe as cosmic consciousness. He describes losing track of his own identity and seeing the "mystic ocean" beneath him as "the visible image of that deep, blue, bottomless soul, pervading mankind and nature; and every strange, half-seen, gliding, beautiful thing that eludes him." Ishmael confesses that in these reveries he fails to keep an eye out for whales as diligently as he's supposed to. "With the problem of the universe revolving in me," he says, "how could I—being left completely to myself at such a thought-engendering altitude." Shipowners, he adds, should beware of hiring the likes of him if they want to make a profit: "Your whales must be seen before they can be killed, and this sunken-eyed young Platonist will tow you ten wakes round the world, and never make you one pint of [whale oil] the richer" (171–73).

Chapter 11. Dreamworld

1. Jaron Lanier, "One Half a Manifesto," *Edge.org*, November 2000, http://bit.ly/147weaJ, accessed May 14, 2013.

2. Ray Kurzweil, *The Singularity Is Near: When Humans Transcend Biology* (New York: Penguin, 2005), 9, 314–15.

3. Zuckerberg's comments on the Oculus VR acquisition are available at https://www .facebook.com/zuck/posts/10101319050523971. Iribe's comments reported by Glen Chapman, "Oculus Out to Let People Touch Virtual Worlds," *Phys.Org.*, June 18, 2015, http:// tinyurl.com/jpog32k. Both accessed September 26, 2015.

4. Matthew Lombard and Theresa Ditton, "At the Heart of It All: The Concept of Presence," *Journal of Computer-Mediated Communication* 3, no. 2 (September 1997), http://tinyurl .com/jpr6z77, accessed March 7, 2016.

5. Ibid.

6. Anne Foerst, *God in the Machine: What Robots Teach Us about Humanity and God* (New York: Dutton, 2004), 9–10; Sherry Turkle, *Life on the Screen: Identity in the Age of the Internet* (New York: Touchstone, 1997), 266.

7. Foerst, *God in the Machine*, 5.

8. Jacob Arlow, "Fantasy, Memory, and Reality Testing," *Psychoanalytic Quarterly* 38 (1969): 29–30. Freud, from *Civilization and Its Discontent*s, quoted by Zygmunt Bauman, *Liquid Times* (Cambridge: Polity Press, 2007), 55–56.

9. Bryon Reeves and Clifford Nass, *The Media Equation: How People Treat Computers, Television and New Media like Real People and Places* (Cambridge: Cambridge University Press, 1996), 7.

10. Joseph Weizenbaum, *Computer Power and Human Reason: From Judgment to Calculation* (San Francisco: W. H. Freeman, 1976), 3–7, 188–91. This passage is based in part on Sherry Turkle's detailed description of the ELIZA controversy. See *Life on the Screen*, 105–14.

11. Dewey, "Psychology and Philosophic Method," lecture delivered at the University of California, May 15, 1899, later reprinted under the title "Consciousness and Experience" in John Dewey, *The Influence of Darwin on Philosophy, and Other Essays in Contemporary Thought* (New York: Henry Holt, 1910), 267. Bush administration official quoted by Ron Suskind, "Faith, Certainty and the Presidency of George W. Bush," *New York Times Magazine*, October 17, 2004, http://tinyurl.com/pysrhzl, accessed November 21, 2015.

12. Jacques Ellul, *The Technological Society* (New York: Vintage, 1964), 372; Postman, *Amusing Ourselves to Death* (1985; repr., New York: Penguin, 2005), 107–8. See also Ellul's *Propaganda: The Formation of Men's Attitudes* (New York: Vintage, 1962), Susan Jacoby's *The Age of American Unreason* (New York: Pantheon, 2008), Frank Rich's *The Greatest Story Ever Sold: The Decline and Fall of Truth from 9/11 to Katrina* (New York: Penguin, 2006), and Joe McGinniss's *The Selling of the President 1968* (New York: Trident Press, 1969).

    The pattern of disconnection from verifiable reality showed no sign of abating as the 2016 presidential campaign got under way. The spectacle of the second Republican debate prompted the *New York Times* to editorialize, "It felt at times as if the speakers were no longer living in a fact-based world," while *Times* columnist Paul Krugman wrote that "the men and woman on that stage are clearly living in a world of fantasies and fictions." See "Crazy Talk at the Republican Debate," September 17, 2015, and "Fantasies and Fictions at G.O.P. Debate," September 18, 2015.

13. Postman, *Amusing Ourselves to Death*, xix–xx.

14. Ibid., xx.

15. Details of the Great Minster Cathedral from Charles Garside, *Zwingli and the Arts* (New Haven, Conn.: Yale University Press, 1966), 87–89.

16. Zwingli quoted by John Dillenberger, "The Seductive Power of the Visual Arts: Shall the Response Be Iconoclasm or Baptism?," *Andover Newton Quarterly* 69, no. 4 (March 1977): 305.

17. On the theology of Orthodox icons, see Rev. Brendan McAnerney, OP, "The Latin West and the Byzantine Tradition of Icons," *Sacred Art Journal*, no. 13–14 (March 1993): 146.

18. Rosalind H. Williams, *Dream Worlds: Mass Consumption in Late Nineteenth-Century France* (Berkeley: University of California Press, 1982), 12, 65–66. On the seductive and educational qualities of department stores, see Alan Trachtenberg, *The Incorporation of America: Culture and Society in The Gilded Age* (1982; repr., New York: Hill and Wang, 2007), 130.

19. William Hewlett and David Packard began their business partnership in 1938, founding a company to make and sell audio oscillators. Their first customer was Walt Disney, who used the devices to prepare theaters to properly play the soundtrack of an ambitious animated film he was producing, which featured classical music. Thus, as Langdon Winner points out, "Silicon Valley literally began with *Fantasia.*" See "Silicon Valley Mystery House," in *Variations on a Theme Park: The New American City and the End of Public Space*, ed. Michael Sorkin (New York: Hill and Wang, 1992), 38.

20. Marshall McLuhan, *Understanding Media: The Extensions of Man* (1964; repr., Cambridge, Mass.: MIT Press, 1998), 26.

21. Turkle, *Life on the Screen*, 13.

22. Ibid., 242.
    It's interesting to note that in 2012, Turkle found it necessary to apologize for what she saw as her failure to appreciate in her early books (*The Second Self: Computers and the Human Spirit* as well as *Life on the Screen*) the dangers posed by the online fantasy games she described. These remarks were made in connection with the publication of her latest book at the time, *Alone Together*, which has become one of the better-known arguments against the distancing effects of social media (arguments that in turn have made Turkle a favorite target of the digital dualism theorists discussed in chapter 9). I found these comments surprising, since from my perspective anyone who reads *Life on the Screen* couldn't help but see that the psychological influences the book describes could hardly be considered wholly benign.

23. T. J. Jackson Lears, *No Place of Grace: Antimodernism and the Transformation of American Culture 1880–1920* (New York: Pantheon Books, 1981), 171.

24. Neil Gabler, *Life: The Movie: How Entertainment Conquered Reality* (New York: Knopf, 1998), 50.

25. Donna Haraway, "A Cyborg Manifesto: Science, Technology, and Socialist-Feminism in the Late Twentieth Century," in *Simians, Cyborgs and Women: The Reinvention of Nature* (New York: Routledge, 1991), 152.

Chapter 12. Abstraction

1. John Lachs, *Responsibility and the Individual in Modern Society* (Brighton, England: Harvester Press, 1981), 13, 57. Zygmunt Bauman explores the contributions of mediation to Germany's prosecution of the Holocaust. See *Modernity and the Holocaust* (1989; repr., Ithaca, N.Y.: Cornell University Press, 2000), 24–26.

2. The meaning I intend for the word "nature" here obviously differs from that intended

in chapter 7. Arthur Lovejoy's taxonomy of the word includes the following: "'Nature' as antithetic to man and his works; the part of empirical reality which has not been transformed (or corrupted) by human art." That is the usage employed in this case. See "'Nature' as Aesthetic Norm," in *Essays in the History of Ideas* (Baltimore: Johns Hopkins University Press, 1948), 71.

3. Steve Lohr, "The Age of Big Data," *New York Times*, February 11, 2012, http://bit.ly /15Z3F2z, accessed August 23, 2013.

4. Emerson quoted by John F. Kasson, *Civilizing the Machine: Technology and Republican Values in America, 1776–1900* (New York: Hill and Wang, 1976), 116. Victor Hugo quoted by Wolfgang Schivelbusch, *The Railway Journey: Trains and Travel in the 19th Century* (Berkeley: University of California Press, 1986), 55. Similar impressions were recorded by many lesser-known writers. See Myron F. Brightfield, "The Coming of the Railroad to Victorian England, as Viewed by Novels of the Period (1840–1870)," *Technology and Culture* 3, no. 1 (Winter 1962): 52–54. Mondrian quoted by Thomas J. Misa, *Leonardo to the Internet* (Baltimore: Johns Hopkins University Press, 2011), 173.

It's fascinating to note how perceptions of the railroad changed over the course of a century. As David E. Nye points out, by the 1950s artists were portraying it in far more bleak and stagnant terms. He cites as an example Edward Hopper's "Hotel by a Railroad," which depicts the relationship between human beings and the railroad as "an emotional and physical cul-de-sac." *America as Second Creation: Technology and Narratives of New Beginnings* (Cambridge: MIT Press, 2003), 203.

5. Oliver Wendell Holmes, "The Stereoscope and the Stereograph," *Atlantic Monthly*, June 1859. This was one of three essays on the new photographic techniques published by Holmes in the *Atlantic*, all published anonymously. See Robert J. Silverman, "The Stereoscope and Photographic Depiction in the 19th Century," *Technology and Culture* 34, no. 4 (October 1993): 736–38.

6. Nietzsche quoted by Maggie Jackson, *Distracted: The Erosion of Attention and the Coming Dark Age* (Amherst, N.Y.: Prometheus Books, 2008), 39. On Nietzsche's comment regarding weightlessness and on Baudelaire, see T. J. Jackson Lears, *No Place of Grace: Antimodernism and the Transformation of American Culture 1880–1920* (New York: Pantheon Books, 1981), 32–33. On Marx, see Marshall Berman's *All That Is Solid Melts into Air: The Experience of Modernity* (New York: Penguin, 1988), 144–46. Berman's book is one of the very best examinations of the psychological and emotional effects of technological change.

7. Israel Kleiner, "Rigor and Proof in Mathematics: A Historical Perspective," *Mathematics Magazine* 64, no. 5 (December 1991), 304–5.

Note that the movement in mathematics toward ever-greater abstraction continued, as evidenced by the contributions of the renowned Russian mathematician Alexander Grothendieck. According to an essay in a journal of the American Mathematics Society, Grothendieck "had an extremely powerful, almost otherworldly ability of abstraction that allowed him to see problems in a highly general context, and he used this ability with exquisite precision. Indeed, the trend toward increased generality and abstraction, which can be seen across the whole field since the middle of the 20th century, is due in no small part to Grothendieck's influence." Quoted in Grothendieck's obituary in the *New York Times*, November 14, 2014, http://tinyurl.com/oly49al, accessed September 12, 2015.

8. Kleiner, "Rigor and Proof in Mathematics," 304–5.

9. Walter Isaacson, *Einstein: His Life and Universe* (New York: Simon and Schuster, 2007), 511–12. See also Peter Galison, "Einstein's Clocks: The Place of Time," *Critical Inquiry* 26, no. 2 (Winter 2000): 355–89.

10. David F. Noble, *The Religion of Technology* (New York: Alfred A. Knopf, 1997), 144–45. See also the biography of Descartes on "The Story of Mathematics" website, http://www.storyofmathematics.com/17th_descartes.html, accessed December 29, 2015.

11. Lovelace quoted by Walter Isaacson, *The Innovators: How a Group of Hackers, Geniuses, and Geeks Created the Digital Revolution* (New York: Simon and Schuster, 2014), 18. Boole quoted by Kleiner, "Rigor and Proof in Mathematics," 302. The quote is from Boole's "The Mathematical Analysis of Logic."

12. Shannon's "A Mathematical Theory of Communications" was published in two parts in the *Bell System Technical Journal* 27 (July and October 1948): 379–423, 623–56.

13. Progress and Freedom Foundation, "Cyberspace and the American Dream: A Magna Carta for the Knowledge Age," 1994, http://www.pff.org/issues-pubs/futureinsights/fi1.2magnacarta.html; John Perry Barlow, "Declaration of Independence of Cyberspace," 1996, https://projects.eff.org/~barlow/Declaration-Final.html, both accessed September 5, 2015.

14. Barlow, "Declaration of Independence of Cyberspace." Note that over time Barlow seems to have wearied of being identified with the sentiments expressed in his cyberspace manifesto. A list of biographical highlights on his personal website ends with this comment: "Finally, he recognizes that there is a difference between information and experience and he vastly prefers the latter."

15. Brockman's statement is from his introduction to the online transcript of an event he hosted at Eastover Farm in Connecticut, "Rebooting Civilization II," in July 2002, https://www.edge.org/events/rebooting-civilzation-ii, accessed May 15, 2016.

16. https://www.edge.org/events/seminars.

17. *Edge*, "Life: A Gene-Centric View," http://bit.ly/18PX4XT, accessed September 9, 2013.

18. Ibid.

19. Ibid.

20. Ray Kurzweil, *The Singularity Is Near: When Humans Transcend Biology* (New York: Penguin, 2005), 132.

21. Georges Canguilhem, "The Role of Analogies and Models in Biological Discovery," in *Scientific Change: Historical Studies in the Intellectual, Social, and Technical Conditions for Scientific Discovery and Technical Invention from Antiquity to the Present*, ed. A. C. Crombie (New York: Basic Books, 1963), 515.

22. Korzybski, *Science and Sanity: An Introduction to Non-Aristotelian Systems and General Semantics* (1st ed., 1933; 5th ed., New York: Institute of General Semantics, Brooklyn, 1994), 58. Lohr, "Age of Big Data." Student quoted by Sherry Turkle, *Simulation and Its Discontents* (Cambridge, Mass.: MIT Press, 2009), 14. Turkle does not identify the student's gender.

Alfred Korzybski's distrust of abstraction went much further than maps to include anything we perceive with our senses and subsequently attempt to record or report, in language and otherwise. His "Principle of Non-Allness," said the media ecology scholar Lance Strate, "reminds us that our perception and knowledge about any given event or object is necessarily incomplete, and all the more so our depictions and descriptions." Strate also quotes Susanne Langer's observation that "the abstractions made by the ear and the eye" are "genuine symbolic materials, media of understanding, by whose office we apprehend a world of *things*." See Strate, *On the Binding Biases of Time* (Fort Worth: Institute of General Semantics, 2011), 24, 32, 131.

23. Other penetrating books on these and related developments include Roger Lowenstein's *The End of Wall Street* (New York: Penguin, 2010), Andrew Ross Sorkin's *Too Big to Fail* (New York: Viking, 2009), Bethany McLean and Peter Elkind's *The Smartest Guys in the Room* (New York: Portfolio, 2005), and Michael Lewis's *Flash Boys: A Wall Street Revolt* (New York: Norton, 2014). See also "How Did Economists Get It So Wrong?," Paul Krugman's analysis of the crisis for the *New York Times Magazine*, September 6, 2009. "As I see it," Krugman wrote, "the economics profession went astray because economists, as a group, mistook beauty, clad in impressive-looking mathematics, for truth." Available at http://www.nytimes.com/2009/09/06/magazine/06Economic-t.html, accessed September 12, 2015.

24. Sherry Turkle, *Life on the Screen: Identity in the Age of the Internet* (New York: Touchstone, 1997), 23, 70–71.

25. Ellen Ullman, *Close to the Machine: Technophilia and Its Discontents* (San Francisco: City Lights Books, 1997).

26. Ibid., 11–12.

27. Ibid., 14–15.

28. Ibid., 89.

29. Abraham Verghese, "Treat the Patient, Not the CT Scan," *New York Times*, February 26, 2011, http://nyti.ms/1c1wmiT, accessed August 23, 2013. In her book *Every Patient Tells a Story: Medical Mysteries and the Art of Diagnosis* (New York: Harmony, 2009), Lisa Sanders also speaks out for the disappearing art of physical examination. See also S. Famani et al., "A Pilot Study Assessing Knowledge of Clinical Signs and Physical Examination Skills in Incoming Medicine Residents," *Journal of Graduate Medical Education*, June 2010, 232–35. "Physical exam skills of medical trainees are declining," the study says, "but most residencies do not offer systematic clinical skills teaching or assessment. . . . Overall, physical exam knowledge and performance of new residents were unsatisfactory" (232).

It's interesting to note that despite the enthusiasm expressed by Oliver Wendell Holmes for the matter-replacing powers of the new techniques of photography, his early reputation as a medical thinker had to some extent been secured by a prize-winning essay he'd written on the importance of direct physical examination of patients. "On the Utility and Importance of Direct Exploration in Medical Practice" focused in particular on the techniques of percussion and auscultation (listening to the sounds of the body, either directly or through a device such as a stethoscope). In the essay Holmes stated that "the time is not far distant, when the physician, who is unable to practice percussion and auscultation, will be held to be unfit for his profession." See Neille Shoemaker, "The Contemporaneous Medical Reputation of Oliver Wendell Holmes," *New England Quarterly* 26, no. 4 (December 1953): 478–79.

30. David S. Cloud, "Anatomy of an Afghan War Tragedy," *Los Angeles Times*, April 10, 2011, http://lat.ms/eKPFiQ. See also Walter Pincus, "Are Drones a Technological Tipping Point in Warfare?," *Washington Post*, April 24, 2011, http://bit.ly/13F5Kyk. Both articles accessed August 4, 2013. Pincus reports on a study by the British Defense Ministry that questioned whether the growing use of armed drones will lead future decision makers (in the words of the study) "[to] resort to war as a policy option far sooner than previously." John McDermott's 1969 essay, "Technology: The Opiate of the Masses," in *Philoso-*

phy of Technology: The Technological Condition, an Anthology, ed. Robert C. Scharff and Val Dusek (Malden, Mass.: Blackwell, 2003), contains a blistering assessment of the use of computer models by the U.S. military in Vietnam.

31. James Gleick, *The Information: A History, a Theory, a Flood* (New York: Pantheon Books, 2011), 8.

Kevin Kelly has argued that whether information theory is metaphor or reality doesn't matter. When the "metaphor of computation" infiltrates physics and biology deeply enough, he says, "It's the metaphor that wins." Kelly first made this comment in the *Whole Earth Review* in 1998 and repeated it in 2001 during a spirited online debate with the philosopher of technology Steve Talbott (author of *The Future Does Not Compute: Transcending the Machines in Our Midst* and *In the Belly of the Beast: Technology, Nature, and the Human Prospect*). The debate touches on many of the issues addressed in this chapter and is well worth reading. It ran in Talbott's online newsletter, *NetFuture*, no. 126 (December 18, 2001), http://bit.ly/1klYxAD, and no. 130 (April 2, 2002), http://bit .ly/1eColP7, both accessed January 6, 2014.

32. The physicist and historian Gerald Holton called the belief that there must be one or a few fundamental elements underlying all creation "the Ionian Enchantment." Arthur Lovejoy referred to this conviction as "uniformitarianism."

33. Julian Offray de la Mettrie, *L'Homme machine (Man a Machine)* is excerpted in *The Philosophy of the Body: Rejections of Cartesian Dualism*, ed. Stuart F. Spicker (Chicago: Quadrangle Books, 1970), 84.

34. N. Katherine Hayles, *How We Became Posthuman: Virtual Bodies in Cybernetics, Literature, and Informatics* (Chicago: University of Chicago Press, 1999), 18–19.

35. Ibid., 19.

36. Philip Marchand, *Marshall McLuhan: The Medium and the Messenger* (Cambridge, Mass.: MIT Press, 1998), 249.

More recently an essay by Richard Kearny, a professor of philosophy at Boston College, posited that we have entered an age of "excarnation," in which our surrender to various forms of digital virtuality has eclipsed our sense of embodiment, with profound consequences. See "Losing Our Touch," *New York Times*, August 30, 2014, http://tinyurl .com/jlbktmj, accessed September 7, 2015.

37. McLuhan quoted in Marchand, *Marshall McLuhan*, 249.

38. Shannon's essay, "The Bandwagon," is available online at http://bit.ly/19juPGk, accessed June 7, 2013.

39. Ibid.

40. Hayles, *How We Became Posthuman*, 112; Jaron Lanier, "One Half a Manifesto," *Edge.org*, November 2000, http://bit.ly/147weaJ, accessed May 14, 2013.

Chapter 13. Shapers Shaped

1. Quoted by Hava Tirosh-Samuelson, "Facing the Challenges of Transhumanism: Philosophical, Religious, and Ethical Considerations," *Global Spiral*, Metanexus Institute, May 10, 2007, http://tinyurl.com/h8kx87h, accessed December 29, 2015.

It would be unfair to conclude from this quote that Nick Bostrom embraces the technological project without reservation. In fact, he believes human extinction is a much more likely outcome than we think, and that technology gone wrong is the main reason why. The advance of transhumanism, he believes, represents one of our better hopes for successfully finding a way out of that dilemma. See Ross Andersen, "We're Underestimating the Risk of Human Extinction," *The Atlantic.com*, March 6, 2012, http://bit.ly /yUUk3G, accessed August 23, 2013.

2. Ray Kurzweil, *The Singularity Is Near: When Humans Transcend Biology* (New York: Penguin, 2005), 175, 127. Minsky quoted by David F. Noble, *The Religion of Technology* (New York: Alfred A. Knopf, 1997), 156.

    I have no idea how many, if any, of the contemporary transhumanists consider themselves Christians, but it's interesting to note in this context Lewis Mumford's argument that the Christian church's traditional distrust of the human body and its desires helped create a cultural opening for the machine that would replace it. Lewis Mumford, *Technics and Civilization* (San Diego: Harcourt Brace, 1934), 35–36.

3. On neuroplasticity, see Jeffrey M. Schwartz and Sharon Begley, *The Mind and the Brain: Neuroplasticity and the Power of Mental Force* (New York: ReganBooks, 2002). Other sources include Gary Small and Gigi Vorgan, *iBrain: Surviving the Technological Alteration of the Modern Mind* (New York: Morrow, 2008), and Nicholas Carr, "Is Google Making Us Stupid?," *Atlantic Monthly*, July/August 2008, http://bit.ly/PBPnQz, accessed August 4, 2013. Carr later expanded this article into a book, *The Shallows: What the Internet Is Doing to Our Brains* (New York: Norton, 2010).

4. Matt Ridley, "What Makes You Who You Are," *Time*, June 2, 2003. Also see John Cloud, "Why Your DNA Isn't Your Destiny," *Time*, January 18, 2010, and Sharon Begley, "Sins of the Grandfathers," *Newsweek*, October 30, 2010.

    The field of study that examines environmental influences on DNA expression is called epigenetics. For a longer examination of its discoveries and their implications, see Julie Gutman and Becky Mansfield, "Plastic People," *Aeon*, February 2013, http:// aeon.co/magazine/science/have-we-drawn-the-wrong-lessons-from-epigenetics/, accessed September 12, 2015.

5. Candice Pert, *Molecules of Emotion: Why You Feel the Way You Feel* (New York: Scribner, 1997), 26–29, 146–48. The molecules discussed by Pert play a role in the enteric nervous system, or ENT. Often called the "second brain," the ENT has become the focus of a relatively new scientific field, "neurogastroenterology." For an overview of the ENT, see "Alimentary Thinking," by Emma Young, *New Scientist*, December 15, 2012, 39–42.

6. Alfred North Whitehead, *Modes of Thought* (New York: Free Press, 1966), 138. Relationality has played an important role in the work of such existential and phenomenologist philosophers as Heidegger, Sartre, and Merleau-Ponty. David Bohm is a physicist who has explored the philosophical and religious implications of quantum physics; see *Wholeness and the Implicate Order* (London: Routledge, 1980).

7. For an overview of the chemical "burden" present in human bodies, see the website of the "Coming Clean" network, http://bit.ly/15sWeY, accessed March 31, 2011. Regarding babies being born pre-polluted, see Ian Urbina, "Think Those Chemicals Have been Tested?," *New York Times*, April 13, 2013, http://nyti.ms/159SlDB, accessed August 4, 2013.

8. Andy Clark, *Natural Born Cyborgs: Minds, Technologies, and the Future of Human Intelligence* (New York: Oxford University Press, 2003), 6–7. It tells us something about how the world has changed, I think, that as an author Clark can deploy such phrases as "biological skinbags" and "Tools-R-Us" while as a professor holding the Ancient Chair of Logic and Metaphysics at Edinburgh University.

9. Emerson quoted by John F. Kasson, *Civilizing the Machine: Technology and Republican Values in America, 1776–1900* (New York: Hill and Wang, 1976), 127.

10. Information on Kapp from Carl Mitcham, *Thinking through Technology: The Path between Engineering and Philosophy* (Chicago: University of Chicago Press, 1994), 23, and from Arnold Gehlen, "A Philosophical-Anthropological Perspective on Technology," in *Philosophy of Technology: The Technological Condition, an Anthology*, ed. Robert Scharff and Val Dusek (Malden, Mass.: Blackwell, 2003), 213–20. Marshall McLuhan, *Understanding Media: The Extensions of Man* (1964; repr., Cambridge, Mass.: MIT Press, 1998), 90.

11. Coxe's arguments and their significance are explored at length in Leo Marx, *The Machine in the Garden: Technology and the Pastoral Ideal in America* (New York: Oxford University Press, 1964), 150–62, from which this quotation is taken. The emphasis is added. See also Jacob E. Cooke, "Tench Coxe, Alexander Hamilton, and the Encouragement of American Manufactures," *William and Mary Quarterly* 32, no. 3 (July 1975), 369–92.

    Both Marx and Cooke cite contemporary evidence that Coxe's character was not entirely trustworthy. According to Marx, John Quincy Adams called him a "wily, winding, subtle and insidious character" (151).

12. Perry Miller, *The Life of the Mind in America: From the Revolution to the Civil War* (New York: Harcourt, Brace, 1965), 304–6. See also David E. Nye, *American Technological Sublime* (Cambridge, Mass.: MIT Press, 1999), 38–39. Writes Nye: "Nature was understood to have authored the script sanctioning its own transformation in the service of an inevitable destiny" (38).

13. Jacques Ellul, *The Technological Society* (New York: Vintage, 1964), 6, 320, 431.

14. Carlyle, "Signs of the Times," *Edinburgh Review*, 1829, the Victorian Web at http://www.victorianweb.org/authors/carlyle/signs1.html, accessed March 16, 2016.

15. Emerson, "Ode (Inscribed to William H. Channing)," in *The Selected Writings of Ralph Waldo Emerson* (New York: Modern Library, 1950), 770; Henry David Thoreau, *Walden* (New York: Thomas Y. Crowell, 1910), 47; Karl Marx, *Capital* (New York: Modern Library, 1906), 421.

16. According to historian George Basalla, Chaplin conceived of the idea for *Modern Times* after a reporter described to him the working conditions in factories in Detroit. "He was told that healthy, young farm boys were lured to the city to work on the assembly line producing automobiles," Basalla writes. "Within four or five years these same young men were 'nervous wrecks,' their health destroyed by the pace of work set in the factory." See "Keaton and Chaplin: The Silent Film's Response to Technology," in *Technology in America: A History of Individuals and Ideas*, 2nd ed., ed. Carroll W. Pursell Jr. (Cambridge, Mass.: MIT Press, 1996), 228, 231–32.

17. Ford quoted by David E. Nye, *Henry Ford: Ignorant Idealist* (Port Washington, N.Y.: Kennikat Press, 1979), 77.

Chapter 14. Ecotone

1. Arnold I. Davidson has written a fascinating essay on the history of taboos associated with bestiality, birth defects, and other perceived violations of the human-animal boundary. See "The Horror of Monsters," in *The Boundaries of Humanity: Humans, Animals, and Machines*, ed. James J. Sheehan and Morton Sosna (Berkeley: University of California Press, 1991), 36–67.

    Several other essays in this collection are relevant to the present discussion. See in particular Harriet Ritvo, "The Animal Connection," and Evelyn Fox Keller, "Language and Ideology in Evolutionary Theory: Reading Cultural Norms into Natural Law."

2. Corbey, *The Metaphysics of Apes: Negotiating the Animal-Human Boundary* (Cambridge: Cambridge University Press, 2005), 21.

It's noteworthy that the philosopher Donna Haraway has followed up her influential essay "The Cyborg Manifesto" with two works on the relationship between animals and humans, *The Companion Species Manifesto: Dogs, People, and Significant Otherness* (Chicago: Prickly Paradigm Press, 2003) and *When Species Meet* (Minneapolis: University of Minnesota Press, 2007). Other books exploring that relationship include Meg Daley Olmert's *Made for Each Other: The Biology of the Human-Animal Bond* (Cambridge, Mass.: Da Capo Press, 2009); Wayne Pacelle's *The Bond: Our Kinship with Animals, Our Call to Defend Them* (New York: Morrow, 2011); Charles Siebert's *The Wauchula Woods Accord: Toward a New Understanding of Animals* (New York: Scribner, 2009); and James Serpell's *In the Company of Animals: A Study of Human-Animal Relationships* (Cambridge: Cambridge University Press, 1996). See also Benedict Cary, "Emotional Power Broker of the Modern Family," *New York Times*, March 14, 2011, http://nyti.ms/unAukr, accessed August 4, 2013.

3. The Cambridge Declaration is available at the Francis Crick Memorial Conference site, http://fcmconference.org/img/CambridgeDeclarationOnConsciousness.pdf, accessed December 29, 2015.

4. Lee M. Silver, "Raising Beast People," *Newsweek*, July 31, 2006. Also see Frans de Waal's *The Age of Empathy: Nature's Lessons for a Kinder Society* (New York: Crown, 2009) and Irene M. Pepperberg's *Alex and Me: How a Scientist and a Parrot Discovered a Hidden World of Animal Intelligence—and Formed a Deep Bond in the Process* (New York: Harper, 2008). For a shorter essay on the subject, see Waal, "The Brains of Animals," *Wall Street Journal*, March 22, 2013, http://on.wsj.com/Z9yyiH, accessed August 4, 2013.

Science writer Jon Cohen has argued that it's time to shift the discussion back in the other direction, emphasizing the differences between ourselves and the animals, specifically chimpanzees, rather than the similarities. See *Almost Chimpanzee: Searching for What Makes Us Human, in Rainforests, Labs, Sanctuaries, and Zoos* (New York: Times Books, 2000).

5. Zackary Canepari and Drea Cooper, "The Family Dog," *New York Times*, June 17, 2015, http://tinyurl.com/ojcu37v, accessed September 27, 2015.

6. Zackary Canepari, Drea Cooper, and Emma Cott, "The Uncanny Lover," *New York Times*, June 11, 2015, http://tinyurl.com/q9mbgph, accessed September 27, 2015.

7. An essay of mine on "creepy" technologies was published by the technology blog *Cyborgology*, http://bit.ly/1dWHJYv, accessed August 4, 2013.

8. George Dyson, *Darwin among the Machines: The Evolution of Global Intelligence* (Cambridge, Mass.: Perseus, 1997), xii. The phrase *homo faber* describes the idea that tool making is what distinguishes human beings from animals. For classic expositions of the *homo faber* theory, see Kenneth B. Oakley, *Man the Tool-Maker* (Chicago: University of Chicago Press, 1949, 1964), and Sherwood L. Washburn, "Tools and Human Evolution," *Scientific American* 203, no. 3 (September 1960): 63–75. See also Hannah Arendt, *The Human Condition* (Chicago: University of Chicago Press, 1958), chap. 4. For a critique of the *homo faber* argument, see Mumford, "Tools and the Man," *Technology and Culture* 1, no. 4 (Autumn 1960): 323–24.

An alternative argument to the *homo faber* theory, originating with the ancient Greeks, holds that the development of language rather than tools represents the demarcation point between human beings and animals. As Marshall McLuhan has described the idea, "Man is distinguished from the brutes by speech, and as he becomes more eloquent he becomes less brutish. As he becomes less brutish he becomes more wise." See Marshall McLuhan, "An Ancient Quarrel in Modern America," *Classical Journal* 41, no. 4 (January 1946): 158.

9. See Bruce Mazlish, *The Fourth Discontinuity: The Co-Evolution of Humans and Machines* (New Haven, Conn.: Yale University Press, 1993). For an earlier, shorter exposition, see Mazlish, "The Fourth Discontinuity," *Technology and Culture* 8, no. 1 (January 1967): 1–15. Glen Mazis has explored similar themes in *Humans, Animals, Machines: Blurring Boundaries* (Albany: SUNY Press, 2008).

10. These comments are based on Tillich's essay, "The Technical City as Symbol," in *The Spiritual Situation in Our Technical Society*, ed. J. Mark Harris (Macon, Ga.: Mercer University Press, 1988), 179–84. In addition to the essays in this superb collection, Tillich also had much to say relevant to technology in the three volumes of his *Systematic Theology*.

11. Ibid., 182.

12. Quoted by Adam Kirsch, "The Redemption of Walter Benjamin," *New York Review of Books*, July 10, 2014.

13. Thomas Hobbes, *Leviathan* (New York: Penguin, 1985), 81.

14. This paragraph is based almost entirely on the work of Herbert L. Sussman and Raymond Corbey. See Herbert L. Sussman, *Victorians and the Machine: The Literary Response to Technology* (Cambridge, Mass.: Harvard University Press, 1968), 135–38, and Corbey, *Metaphysics of Apes*, 70–74.

15. Charles Darwin, *The Descent of Man* (1874; Amherst, N.Y.: Prometheus Books, 1998), 643. The contemporary songwriter Nick Lowe is among those who has recognized within himself, as Darwin put it, "the indelible stamp of his lowly origin." See the lyrics to "The Beast in Me." Bruce Springsteen has made a similar confession; see the lyrics to "Part Man, Part Monkey."

16. On Huxley's reconsiderations, see Michael Ruse, "The Darwinian Revolution: Rethinking Its Meaning and Significance," *Proceedings of the National Academy of Sciences*, 106, Suppl. 1 (June 16, 2009): 10044, http://1.usa.gov/1ekSuEl, accessed August 4, 2013. Sigmund Freud, *Civilization and Its Discontents*, trans. James Strachey (New York: W. W. Norton, 1961; 1st German ed., 1930), 38–39.

17. I first heard the term "ecotone" when the TV series *Six Feet Under* used it as the title of an episode. The episode opened with a hiker in the hills outside Los Angeles being attacked and eaten by a mountain lion.

Chapter 15. Gamblers

1. Norbert Wiener, *God and Golem, Inc.: A Comment on Certain Points Where Cybernetics Impinges on Religion* (Cambridge, Mass.: MIT Press, 1964), 69.

2. Norbert Wiener, *The Human Use of Human Beings: Cybernetics and Society* (Boston: Houghton Mifflin, 1950), 16. My 2013 essay, "Hello Robots, Goodbye Fry Cooks," contains links to a number of articles on the threat that advances in automation pose to jobs. It's available at my blog, http://tinyurl.com/bc9hjp6. See also the 2015 book by Martin Ford, *Rise of the Robots: Technology and the Threat of a Jobless Future* (New York: Basic Books, 2015).

3. Wiener, *Human Use of Human Beings*, 41.

4. In their superb biography of Wiener, Flo Conway and Jim Siegelman document the campaign of Wiener's father to publicize his son's prodigious intellectual gifts, efforts that had significant consequences for Wiener's psychological well-being. See *Dark Hero of the Information Age: In Search of Norbert Wiener, the Father of Cybernetics* (New York: Basic Books, 2005), part 1.

5. Conway and Siegelman, *Dark Hero*, 16. My article "Norbert Wiener and the Counter-Tradition to the Dream of Mastery" discusses Wiener's lifelong appreciation of uncertainty. *IEEE Technology and Society Magazine*, September 2015, http://tinyurl.com/jxw2ewa.

6. Wiener, *God and Golem, Inc.*, 53. On technological gamblers, see Joseph Weizenbaum, *Computer Power and Human Reason: From Judgment to Calculation* (San Francisco: W. H. Freeman, 1976), 121–27. Note also the title of Jacques Ellul's *The Technological Bluff* (Grand Rapids, Mich.: Wm. B. Eerdmans, 1990).

7. Rob Carlson, *Biology Is Technology: The Promise, Peril, and New Business of Engineering Life* (Cambridge, Mass.: Harvard University Press, 2010), 5.

8. Kevin Kelly, *Out of Control: The New Biology of Machines, Social Systems, and the Economic World* (Cambridge, Mass.: Perseus, 1994), 257. Ray Kurzweil, *The Singularity Is Near: When Humans Transcend Biology* (New York: Penguin, 2005), 45.

   In fairness it should be noted that in his subsequent book, Kelly takes pains to temper his enthusiasm somewhat. Technological change will turn out for the best more than half the time, he says. Even if the average comes in at slightly more than 50 percent, he adds, in the long run that will be good enough. When a given technology proves to be damaging, "we can make better technologies." See Kevin Kelly, *What Technology Wants* (New York: Viking, 2010), 74, 101.

   While more cautious than some of Kelly's earlier writings, this logic nonetheless seems to overlook the fact that if the damage inflicted by a given technology is serious enough, it doesn't really matter that it was an exception to the general rule.

9. For a snapshot of what passes for regulatory oversight of nanotechnology development, see my blog essay on the performance of the National Nanotechnology Initiative, which is responsible for allocating federal funds for nano research. The essay can be found at http://tinyurl.com/86abu23, accessed May 12, 2012.

10. "U.S. Trends in Synthetic Biology Research Funding," Wilson Center, September 15, 2015, http://tinyurl.com/gomoj5v, accessed December 13, 2015.

    In December 2010 the Presidential Commission for the Study of Bioethical Issues published its conclusions on the safety of synthetic biology, saying it had found "no reason to endorse additional federal regulations or a moratorium on work in this field at this time." An open letter signed by more than fifty environmental organizations called that conclusion hopelessly inadequate. "We are disappointed that 'business as usual' has won out over precaution in the commission's report," the letter said. "Self regulation amounts to no regulation." Available online at http://bit.ly/13MHJez, accessed August 23, 2013.

11. Antonio Regalado of *MIT Technology Review* has done a brilliant job of covering the CRISPR story. In December 2015 he published "Everything You Need to Know about Gene Editing's Monster Year," which contains links to several of his articles. Available at http://tinyurl.com/om597q8. For a superb general piece on CRISPR, see Heidi Ledford,

"CRISPR, the Disruptor," *Nature*, June 8, 2015, http://tinyurl.com/joxxprm. Both accessed March 8, 2016.

    On the issue of producing human babies with edited DNA, see two of Antonio Regalado's articles in *MIT Technology Review*: "Engineering the Perfect Baby" and "Scientists Call for a Summit On Gene-Edited Babies." Available at http://tinyurl.com/ktv7e5c and http://tinyurl.com/mk3l4np; both accessed December 29, 2015.

12. See Antonio Regalado, "With This Genetic Engineering Technology, There's No Turning Back," *MIT Technology Review*, November 23, 2015, http://tinyurl.com/ok76pgn, accessed December 29, 2015.

13. See Ledford, "CRISPR, the Disruptor."

14. See two articles by Antonio Regalado in *MIT Technology Review*, "Protect Society from Our Inventions, Say Genome-Editing Scientists," July 17, 2014, http://tinyurl.com /zfpzx3u; and "Chinese Team Reports Gene-Editing Human Embryos," April 22, 2015, http://tinyurl.com/nphal4a, both accessed December 29, 2015.

15. Antonio Regalado, "Scientists on Gene-Edited Babies: It's 'Irresponsible' for Now," *MIT Technology Review*, December 3, 2015, http://tinyurl.com/jg5qaxy. See also Nicholas Wade, "Scientists Seek Moratorium on Edits to Human Genome That Could Be Inherited," *New York Times*, December 3, 2015, http://tinyurl.com/jdzgqho. Both accessed December 29, 2015.

16. Antonio Regalado, "Patients Favor Changing the Genes of the Next Generation," *MIT Technology Review*, December 2, 2015, http://tinyurl.com/jclsad6, accessed March 8, 2016.

17. See also Amy Harmon, "Open Season Is Seen in Gene Editing of Animals," *New York Times*, November 26, 2015, http://tinyurl.com/nqnnkb7, accessed March 8, 2016. "Statement for the Record, Worldwide Threat Assessment of the U.S. Intelligence Community," presented to the Senate Armed Services Committee by James R. Clapper, director of National Intelligence, on February 9, 2016, http://tinyurl.com/hzds6q9. Several leading geneticists and bioethicists subsequently spoke in support of that assessment. See Alan Yuhas and Kamala Kelkar, "'Rogue Scientists' Could Exploit Gene Editing Technology, Experts Warn," *Guardian*, February 12, 2016, http://tinyurl.com/gs8g2mp. Both accessed March 26, 2016.

18. Bill Joy, "Why the Future Doesn't Need Us," *Wired*, April 2000, http://www.wired.com /2000/04/joy-2/, accessed March 16, 2016.

19. Wiener, *Human Use of Human Beings*, 57.

Chapter 16. Consequences

1. Nathan Bomey, "Takata Airbag Recall Now Largest in U.S. History," *USA Today*, May 4, 2016. Trip Gabriel and Coral Davenport, "Calls for Oversight in West Virginia Went Unheeded," *New York Times*, January 13, 2014, http://tinyurl.com/mgaungn, accessed October 3, 2015.

2. Norbert Wiener, "Some Moral and Technical Consequences of Automation," *Science*, n.s., 131, no. 3410 (May 6, 1960): 1355.

3. Nelson D. Schwartz and Louise Story, "Surge of Computer Selling after Apparent Glitch Sends Stocks Plunging," *New York Times*, May 6, 2010, http://nyti.ms/aAzXcc, accessed March 9, 2016. Andrea Chang and Tracey Lien, "Outages at NYSE, United Airlines, WSJ

.com Expose Digital Vulnerabilities," *Los Angeles Times*, July 8, 2015, http://tinyurl.com /q4xhceo, accessed December 8, 2015.

    Flash Boys: A Wall Street Revolt by Michael Lewis (New York: Norton, 2014) documents how high-speed trading has come to dominate the stock markets, with devastating results for their stability and integrity.

4. "Large Hadron Collider Scuttled by Birdy Baguette-Bomber," *Register*, November 5, 2009, http://bit.ly/1bPHUZ3, accessed April 5, 2011.

5. Robert K. Merton, "The Unanticipated Consequences of Purposive Social Action," *American Sociological Review* 1, no. 6 (December 1936): 894.

6. Langdon Winner, *Autonomous Technology: Technics-out-of-Control as a Theme in Political Thought* (Cambridge, Mass.: MIT Press, 1977), 93–97.

7. Ibid., 93.

8. Ibid., 95, 96–97.

9. Steve J. Heims, *John Von Neumann and Norbert Wiener: From Mathematics to the Technologies of Life and Death* (Cambridge, Mass.: MIT Press, 1980), p. 343.

10. Ibid.

11. Wiener quoted by Flo Conway and Jim Siegelman, *Dark Hero of the Information Age: In Search of Norbert Wiener, the Father of Cybernetics* (New York: Basic Books, 2005), 238.

12. Joy, "Why the Future Doesn't Need Us."

13. This Wilson quote is taken from the second installment of the *Atlantic Monthly's* serialization of *Consilience*, March 1998, http://bit.ly/12l6k9H, accessed July 21, 2011.

14. Michael Specter, "A Life of Its Own," *New Yorker*, September 28, 2009, http://nyr.kr /wPKfw, accessed August 4, 2013.

15. Endy's comments before the Presidential Commission for the Study of Bioethical Issues are available at bioethics.gov. See transcripts and/or videos for two panels on July 8, 2010: "Overview and Context of the Science and Technology of Synthetic Biology" and "Speakers' Roundtable."

16. *Edge.org* interview, February 19, 2008, http://bit.ly/13eRXyk, accessed March 17, 2011.

17. Freeman Dyson, "Our Biotech Future," *New York Review of Books*, July 19, 2007, http://bit .ly/jOdMy2, accessed August 4, 2013.

    In his book *Biopunk: DIY Scientists Hack the Software of Life* (New York: Penguin, 2011), Marcus Wohlsen writes, "Nothing comes closer to a founding text of biohacking" than Dyson's article. Because of it, Wohlsen adds, biohackers consider Dyson a "patron saint."

18. Dyson, "Our Biotech Future."

19. In addition to his numerous contributions to theoretical physics, Dyson has a long history of endorsing daring ideas, among them human habitation of space. As of this writing he serves as the president of the Space Studies Institute, founded by his late Princeton colleague Gerard K. O'Neill (see chapter 5). He is also on the Board of Governors of the National Space Society, as is Eric Drexler. The society, which is the offspring of the L5 Society and the National Space Institute, describes itself as "an independent, educational, grassroots, non-profit organization dedicated to the creation of a spacefaring civilization." Its mission: "People living and working in thriving communities beyond the Earth, and the use of the vast resources of space for the dramatic betterment of humanity." See http://www.nss.org/about/, accessed March 28, 2016.

    In light of my earlier discussion of the space colonies project fronted by Gerard O'Neill, I should note that Dyson's article in the *New York Review of Books* prompted a spirited response from Wendell Berry (September 27, 2007, issue, http://bit.ly/13cNoFc, accessed March 9, 2016). "This of course is only another item," Berry wrote, "in a long wish list of techno-scientific panaceas that includes the 'labor-saving' industrialization of virtually everything, eugenics (the ghost and possibility that haunts genetic engineering), chemistry (for 'better living'), the 'peaceful atom,' the Green Revolution, television, the space program, and computers."

20. For an examination of problems created by the importation of novel species into unfamiliar environments, see Chris Bright's *Life out of Bounds: Bioinvasion in a Borderless World* (New York: Norton, 1998).

21. The Ecomodernist Manifesto is available at http://www.ecomodernism.org. For a response, see "A Call to Look Past *An Ecomodernist Manifesto*: A Degrowth Critique" at http://tinyurl.com/zyaxd3x. Both accessed December 23, 2015.

22. The phrase "supreme emergency" is used by the just-war theorist Michael Walzer to describe one of the conditions that would justify a nation's launching of aggressive action in self-defense against the aggressive act of another nation. I've written an essay describing how the environmental, economic, and social disruptions caused by global warming could, sooner or later, fulfill the requirements prescribed by just war theory as legitimate rationales for such action. I describe the essay as a "thought experiment" because I definitely do not encourage or endorse aggressive action; I simply believe that climate change is driving us in that direction. I suspect that if nations of the world some day decide to initiate aggressive action in response to climate change, it won't be aimed so much at removing the sources of global warming as at countering the *results* of global warming. In other words, I think it likely that nations will go to war to assure the security of their populations and their economies as the disruptions of climate change accumulate. The essay, "Just War Theory and Climate Change," is available at my blog, http://tinyurl.com/hbmfem8, accessed December 23, 2015.

Chapter 17. Hubris

1. Bacon quoted by Douglas Groothuis, "Bacon and Pascal on Mastery over Nature," *Research in Philosophy and Technology*, vol. 14 (Greenwich, Conn.: JAI Press, 1994), 195.

2. Isaiah 14:13–14 (New International Version).

3. *Time* magazine cover story on Steve Jobs, April 1, 2010.

4. As Walter Isaacson put it, Jobs "always believed that the rules that applied to ordinary people didn't apply to him." Quoted by James B. Stewart, "Steve Jobs Defied Convention, and Perhaps the Law," *New York Times*, May 2, 2014, http://tinyurl.com/lmr7867, accessed September 7, 2015.

5. On Neumann favoring a preemptive nuclear strike, see Kai Bird and Martin J. Sherwin, *American Prometheus: The Triumph and Tragedy of J. Robert Oppenheimer* (New York: Vintage, 2006), 378. Steve J. Heims, *John Von Neumann and Norbert Wiener: From Mathematics to the Technologies of Life and Death* (Cambridge, Mass.: MIT Press, 1980), 368–69.

6. John von Neumann, "Can We Survive Technology?," *Fortune*, June 1955. Flo Conway and Jim Siegelman's biography of Wiener, *Dark Hero of the Information Age: In Search of Norbert Wiener, the Father of Cybernetics* (New York: Basic Books, 2005), portrays Neumann in a somewhat villainous light, suggesting, among other things, that he incorporated elements of Wiener's research into his own without credit, and that he misled Wiener into thinking he would join him in collaborative research at MIT while secretly making other arrangements at Princeton.

7. Steve J. Heims, *John Von Neumann and Norbert Wiener*, 369–70. Neumann's close ties to the military were responsible for a compelling scene near the end of his life. As he lay dying, guards were posted at his hospital room to ensure that in his delirium he wouldn't disclose any national security secrets.

8. Mark Dery, *Escape Velocity: Cyberculture at the End of the Century* (New York: Grove Press, 1996), 306–7.

   Moravec's comment is reminiscent of a crucial moment in Goethe's *Faust*. Faust is rushing his grand development project to completion, but an old couple whose cottage sits in the middle of his expansive view doesn't want to move. Faust urges Mephistopheles to get rid of them, explaining with a shrug, "That stubbornness, perverse and vain, / So blights the most majestic gain / That to one's agonized disgust / One has to tire of being just" (Norton Critical Edition, 11269–71).

9. Bacon quoted by David F. Noble, *The Religion of Technology* (New York: Alfred A. Knopf, 1997), 49–51. Brand has acknowledged that he borrowed his famous line from the British anthropologist Edmund Leach.

10. Burke quoted by John Updike, "O Beautiful for Spacious Skies," *New York Review of Books*, August 15, 2002, 26; David E. Nye, *American Technological Sublime* (Cambridge, Mass.: MIT Press, 1999), 16.

11. *North American Review* 33, no. 72 (July 1831): 126. This essay was published anonymously by an author identified by later historians as Timothy Walker.

12. Henry Adams, *The Education of Henry Adams: An Autobiography* (1907; repr., Boston: Houghton Mifflin, 2000), chap. 25. The transition from the natural to the technological sublime can be tracked by comparing Henry Adams's reaction to the dynamo to the reaction of his grandfather, John Quincy Adams, to Niagara Falls: "I have seen it in all its sublimity and glory," the elder Adams wrote, "—and I have never witnessed a scene its equal . . . a feeling over-powering, and which takes away the power of speech by its grandeur and sublimity." Nye, *American Technological Sublime*, 21.

13. Oppenheimer quoted by Nye, *American Technological Sublime*, 228. Norman Mailer, *Of a Fire on the Moon*, in the collection *The Time of Our Time* (New York: Random House, 1998), 737.

14. Langdon Winner, *Autonomous Technology: Technics-out-of-Control as a Theme in Political Thought* (Cambridge, Mass.: MIT Press, 1977), 141; Robert Scharff and Val Dusek, *Philosophy of Technology: The Technological Condition, an Anthology* (Malden, Mass.: Blackwell, 2003), 6.

15. My thoughts on technocracy have been especially informed by William Atkin's *Technocracy and the American Dream: The Technocrat Movement, 1900–1941* (Berkeley: University of California Press, 1977). This section also relies on Edwin Layton Jr., *The Revolt of the Engineers: Social Responsibility and the American Engineering Profession* (Baltimore: Johns Hopkins University Press, 1986); and Henry Elsner Jr., *The Technocrats: Prophets of Automation* (Syracuse, N.Y.: Syracuse University Press, 1967). Langdon Winner's chapter on technocracy in *Autonomous Technology* was also helpful.

16. Statistics on the growth of American engineering graduates are from Thomas Parke Hughes, *American Genesis: A History of the American Genius for Invention* (New York: Penguin, 1989), 243. Edwin Layton Jr. provides statistics that show the same trend from a slightly different perspective. Between 1880 and 1920, he says, the engineering profession as a whole grew from 7,000 to 136,000, an almost twentyfold increase. See "Veblen and the Engineers," *American Quarterly* 14, no. 1 (Spring 1962): 70.

17. Prout quoted by Hugo A. Meier, "American Technology and the Nineteenth-Century World," *American Quarterly* 10, no. 2, part 1 (Summer 1958): 130.

18. The phrase "myth of the engineer" is William Atkin's. The quoted phrases from Veblen are from Winner, *Autonomous Technology*, 144.

19. Howard P. Segal, *Technological Utopianism in American Culture* (Chicago: University of Chicago Press, 1985), 121. Regarding the engineers of the 1950s, Matthew Wisnioski points out that the sociologist C. Wight Mills, in *White Collar: The American Middle Classes* (1951), portrayed engineers as archetypical company men, going along to get along. See *Engineers for Change: Competing Visions of Technology in 1960s America* (Cambridge, Mass.: MIT Press, 2012), 28. For Wisnioski's description of the similar characteristics of most engineers in the 1960s, see my note 11 for chapter 3 and note 1 for chapter 5.

20. For information on the size and characteristics of the Technocracy movement's membership, see Atkin, *Technocracy and the American Dream*, 101–8. Technocracy pamphlet quoted by Segal, *Technological Utopianism in American Culture*, 122.

   The economists Joseph Stiglitz and Linda J. Bilmes have argued that technology was the cause of the Great Depression, rather than its potential savior. Contrary to a popular view that the Depression was the result of misguided federal economic policies, they write, the crisis was set in motion by the massive mechanization of farming during the decades leading up to 1930, together with the introduction of new fertilizers, which simultaneously led to greatly increased productivity achieved with far fewer hands. The result was sharp declines in food prices, sharp declines in farmers' incomes, sharp declines in employment, and sharp declines in consumer spending—a downward economic spiral, in other words. The massive increase in government spending related to World War II, Stiglitz and Bilmes believe, created the bridge that eventually allowed the nation to complete its transformation from an agricultural to an industrial economy. See "The Book of Jobs," *Vanity Fair*, January 2012, http://tinyurl.com/heuxsnt, accessed October 2, 2015.

21. Atkin, *Technocracy and the American Dream*, chap. 9. It's interesting to note, in light of my discussion of the classic/romantic split in chapters 3 and 4, that among the reforms technocrats envisioned as necessary in order to "condition" the masses was an overhaul of the educational system. They proposed eliminating the existing liberal arts curriculum in favor of technical instruction, or, as Atkin put it, replacing humanities classes with "shop" classes.

   And, as they say, the more things change the more they stay the same. On June 13, 2010, an article appeared in the *New York Times* with the headline "Studying Engineering Before They Can Spell It." It described a national trend toward adding engineering courses to elementary school science curricula. The trend was driven, the article said, by "growing concerns that American students lack the skills to compete in a global economy." Also spurring the revised curricula was the lure of billions of dollars in government grants for "STEM" programs. STEM stands for science, technology, engineering, and math. The article added that Congress was considering new legislation to promote engineering education for children from kindergarten through twelfth grade. According to the article, such legislation had been endorsed by more than one hundred businesses and organizations, among them IBM and Lockheed Martin. Article available at http://nyti.ms/aH33V8, accessed November 17, 2015.

22. Ken Auletta, *Googled: The End of the World as We Know It* (New York: Penguin, 2009), 227.

23. See Nicholas Carson, "The Untold Story of Larry Page's Incredible Comeback," *Slate*, April 25, 2014, http://tinyurl.com/n5w529m, accessed September 5, 2015.

    Much the same attitude prevailed at Facebook. When Katherine Losse worked there, she was puzzled that Mark Zuckerberg would invariably open staff meetings with the declaration, "Facebook is a technical company." Eventually she understood that this was a recruiting message Zuckerberg wanted to spread in Silicon Valley. "Good engineers will only work at a company that grants privileges to the technical people," Losse wrote. "They need to know that their ideas and decisions will be considered primary, and not those of marketing or business guys." *The Boy Kings: A Journey into the Heart of the Social Network* (New York: Free Press, 2012), 119–20.

24. Adam Bryant, "Google's Quest to Build a Better Boss," *New York Times*, March 12, 2001, http://nyti.ms/eQvSW2, accessed August 4, 2013.

25. Tom Hamburger and Matea Gold, "Google, Once Disdainful of Lobbying, Now a Master of Washington Influence," *Washington Post*, April 12, 2014, http://tinyurl.com/zuq63l5, accessed October 2, 2015. On Google's efforts in Europe, see Simon Marks and Harry Davies, "Revealed: How Google Enlisted Members of U.S. Congress It Bankrolled to Fight $6bn EU Antitrust Case," *Guardian*, December 17, 2015, http://tinyurl.com/hfsz4jw, accessed December 30, 2015.

    It's interesting to note that the prosecutor in charge of the EU case against Google, Margrethe Vestager of Denmark, has been described as more of a technocrat than a politician. "She's data-driven, and in that sense, unideological," a Danish journalist told the *Washington Post* (a comment that reflects the common belief that being data-driven means you're free of ideology and that fails to entertain the possibility that being data driven might in fact *be* a form of ideology). See Todd C. Frankel, "In E.U. Antitrust Suit, Google Is up against 'a Tough Cookie,'" http://tinyurl.com/z2p733j, accessed December 30, 2015.

26. "Our Story," FWD.us website, http://www.fwd.us/about_us, accessed April 1, 2016.

27. On FWD.us's stumbles, see Jessica Meyers, "FWD.us Immigration Push Hits Wall," *Politico*, July 8, 2014, http://tinyurl.com/htdzhye, accessed April 1, 2016. On Zuckerberg's Newark investment, see Dale Russakoff's *The Prize: Who's in Charge of America's Schools?* (New York: Houghton Mifflin Harcourt, 2015). Reich, *New York Times*, September 18, 2015, http://tinyurl.com/pgjj2tr, accessed December 30, 2015.

28. Contrary to popular belief, President Jimmy Carter was not an engineer, although he came close. After graduating from the United States Naval Academy with a bachelor of science degree, he joined the Navy's nuclear submarine program under Admiral Hyman G. Rickover and studied nuclear power at the Atomic Energy Commission's Naval Reactors Branch. During this period he worked on the USS *Seawolf*, one of the first American nuclear submarines, and in 1952 was dispatched to Deep River, Ontario, to help deal with a meltdown at the nuclear plant there. Carter's apprenticeship as a nuclear engineer was cut short the following year, however, when his father died and he returned to Georgia to take over the family peanut farm. For a timeline of Carter's career, see "Jimmy Carter," *American Experience*, Public Broadcasting System, http://tinyurl.com/j9rg8p5, accessed December 10, 2015.

    Note that some historians have argued that technocracy set the stage for the New Deal, which then co-opted the movement's strategies for reform. The historian William Atkin disagrees. Technology, he said, "had little direct influence on the politics of the New Deal years." See *Technology and the American Dream*, xi, 110–11.

29. Unless otherwise noted, this section is based on David Halberstam, *The Best and the Brightest* (1969; repr., New York: Random House, 1972); and Tim Weiner, "Robert S. McNamara, Architect of a Futile War, Dies at 93," *New York Times*, July 7, 2009, http://nyti .ms/19TMH3, accessed August 4, 2013. Also worth mentioning again in this context is John McDermott's 1969 essay "Technology: The Opiate of the Intellectuals" (available in *Philosophy of Technology: The Technological Condition, an Anthology*, ed. Robert C. Scharff and Val Dusek [Malden, Mass.: Blackwell, 2003]), which examines the technological methodology that the United States assumed would assure victory in Vietnam.

30. Weiner, "Robert S. McNamara."

31. McNamara's book quoted by Theodore Roszak, *The Making of a Counter Culture* (1968; repr., Berkeley: University of California Press, 1995), 11–12.

32. Weiner, "Robert S. McNamara."

33. Halberstam, *Best and the Brightest*, 214.

34. A transcript of the film is available on Errol Morris's website at http://www .errolmorris.com/film/fow_transcript.html, accessed March 19, 2016.

35. Berman, excerpted in the Norton critical edition of *Faust*, p. 727, referring to *Faust*, 11563.

36. Mary Shelley, *Frankenstein* (New York: Bantam, 1991 reissue), 194 (chap. 24).

37. Norbert Wiener, *The Human Use of Human Beings: Cybernetics and Society* (Boston: Houghton Mifflin, 1950), 184.

38. Norbert Wiener, *God and Golem, Inc.: A Comment on Certain Points Where Cybernetics Impinges on Religion* (Cambridge, Mass.: MIT Press, 1964), 56.

Conclusion

1. Perry Miller, *The Responsibility of Mind in a Civilization of Machines*, ed. John Crowell and Stanford J. Searl Jr. (Amherst: University of Massachusetts Press, 1979), 201.

2. Steven Pinker's *The Better Angels of Our Nature: Why Violence Has Declined* (New York: Viking, 2011) contends that we're living in the least violent period in history, but the mixed reviews the book received suggest the evidence leaves room for debate. It's worth noting that, although *The Better Angels of Our Nature* isn't a study of technology—it argues that violence has been reduced by the spread of Enlightenment ideas and attitudes—Pinker does cite technology as "the single most important exogenous cause" of that spread. See p. 477.

INDEX